552357

D1187968

WITHDRAWN FROM LIBRARY STOCK

Renewable Energy in
Power Systems

Renewable Energy in Power Systems

Leon Freris

Visiting Professor in Renewable Energy
Centre for Renewable Energy Systems Technology (CREST),
Loughborough University, UK

David Infield

Professor of Renewable Energy Technologies
Institute of Energy and Environment,
University of Strathclyde, UK

A John Wiley & Sons, Ltd, Publication

This edition first published 2008
© 2008, John Wiley & Sons, Ltd

Registered office
John Wiley & Sons Ltd, The Atrium, Southern Gate, Chichester, West Sussex, PO19 8SQ, United Kingdom

For details of our global editorial offices, for customer services and for information about how to apply for permission to reuse the copyright material in this book please see our website at www.wiley.com.

The right of the author to be identified as the author of this work has been asserted in accordance with the Copyright, Designs and Patents Act 1988.

Reprinted with corrections March 2009.

All rights reserved. No part of this publication may be reproduced, stored in a retrieval system, or transmitted, in any form or by any means, electronic, mechanical, photocopying, recording or otherwise, except as permitted by the UK Copyright, Designs and Patents Act 1988, without the prior permission of the publisher.

Wiley also publishes its books in a variety of electronic formats. Some content that appears in print may not be available in electronic books.

Designations used by companies to distinguish their products are often claimed as trademarks. All brand names and product names used in this book are trade names, service marks, trademarks or registered trademarks of their respective owners. The publisher is not associated with any product or vendor mentioned in this book. This publication is designed to provide accurate and authoritative information in regard to the subject matter covered. It is sold on the understanding that the publisher is not engaged in rendering professional services. If professional advice or other expert assistance is required, the services of a competent professional should be sought.

Library of Congress Cataloging-in-Publication Data

Infield, D. G.
    Renewable energy in power systems / Leon Freris, David Infield.
        p. cm.
    Includes bibliographical references and index.
    ISBN 978-0-470-01749-4 (cloth)
1. Renewable energy sources.   I. Freris, L. L.   II. Title.
    TJ808.I54 2008
    621.4–dc22

                                        2007050173

SOUTHA...
UN... ...Y
SUPPLIE... BLACKW
ORDER ...
DATE   lo .7. 09

A catalogue record for this book is available from the British Library

ISBN 978-0-470-01749-4 (H/B)

Set in 10 on 12 Times by SNP Best-set Typesetter Ltd., Hong Kong
Printed in Great Britain by CPI Antony Rowe, Chippenham, Wiltshire

Cover image © Ted Leeming
Reproduced by permission of Ted Leeming

Contents

Foreword

By Jonathon Porritt

You can read the current state of awareness about climate change any which way you want. You can continue to ignore (or even deny) the overwhelming scientific consensus that has gradually emerged over the last few years. You can get totally lost in the intricacies of climate policy and the political controversies about who is doing or not doing what. But 20 years into the debate about climate change, one thing is overwhelmingly clear: the future of human kind depends in large measure on the speed with which we can massively expand the contribution of renewable energy to our overall energy needs.

That the world is now on a collision course is not seriously disputed. The International Energy Agency constantly reminds people that overall energy use will at least double by 2030 and that most of that expansion will be powered by growth in fossil fuels. On the other hand, climate scientists now tell us that we will need to reduce emissions of CO_2 and other greenhouse gases by at least 60% by 2050. It doesn't remotely begin to add up.

Which makes it hard to understand why so many people are still so crabby and cautious in defining the role for renewables. All their projections are based on 'business-as-usual' economic models – as if any of those are going to be terribly relevant for very much longer.

Indeed, this is the one area where I believe it really is legitimate to talk about 'going onto a war footing' in combating the threat of runaway climate change. And that may not be so far off. For instance, if the price of oil stays at or around \$100 a barrel, and the price of a tonne of CO_2 rises rapidly over the next 3 or 4 years, much of the rubbish still being talked about renewables being 'uneconomic' will just wither away.

That, however, is only the start of it. I have been giving lectures to CREST students for the best part of 10 years, and have learnt during that time that even if the technologies themselves are rapidly improving, and even if the political and economic context could be completely transformed, as I believe is now possible, the real challenge lies in accommodating high penetrations of these new technologies in the electricity supply system, by adapting existing networks and/or the creation of new infrastructure for transmission and distribution. That's where much of the innovation (and huge amounts of new investment) will be needed over the next few years.

And that is one of the greatest strengths of this hugely informative new book: connecting up all the dots so that a clear and utterly convincing picture emerges. And that means taking proper account of the critical importance of energy efficiency (so often ignored in treatments of renewable energy), energy security, and the kind of governance systems which will be needed to drive forward so very different an energy economy.

This is complex, challenging territory, for which reliable and very experienced guides are strongly recommended!

Jonathon Porritt is Founder Director of Forum for the Future www.forumforthefuture.org. uk, Chairman of the UK Sustainable Development Commission www.sd-commission.org.uk, and author of *Capitalism as if the World Matters; Revised Edition 2007* (in paperback), Earthscan – available through 'Forum for the Future' website.

Preface

There is worldwide agreement on the need to reduce greenhouse gas emissions, and different policies are evolving both internationally and locally to achieve this. On 10 January 2007 the EU Commission announced an Energy Package which was endorsed by the European Council. The objectives are that by 2020 EU greenhouse gases are to be reduced by 30 % if a global agreement is arrived at or by 20 % unilaterally. One of the vital components in the achievement of this goal is the intention to provide a 20 % share of energy from renewable energy (RE) sources in the overall EU energy mix.

At present, wind power is the leading source of new renewable energy. World wind power capacity has been growing rapidly at an average cumulative rate of 30 % over the last ten years. About 20 GW of new capacity was installed in 2007 bringing the world total in that year to 94 GW. This annual investment represents around 25 billion euros by an industry that employs 200 000 people and supplies the electricity needs of 25 million households. This considerable expansion has attracted investment from major manufacturing companies such as General Electric, Siemens, ABB and Shell as well as numerous electricity utilities, notably E.ON and Scottish Power. The future of wind power over the next two decades is bright indeed.

Generation of electricity from the sun can be achieved directly using photovoltaic (PV) cells or through solar concentration to raise steam and drive conventional turbines. Over the last few years considerable progress has been made in the reduction of the cost of PV generated electricity, with 2006 seeing the total value of installed capacity reaching 15 billion euros and with cell global production in that year approaching 2.5 GW. It is expected that further technology improvement and production cost reduction over the next decade will result in wide scale competitive generation from this source.

Marine energy is an exciting, but less well developed technology. Tidal barrages, tidal stream turbines and wave energy devices are all in the experimental and pre-commercial stage but are expected to make a significant contribution by around 2015. Geothermal energy is now established in countries like Iceland with a significant accessible resource, and as the technology develops could be taken up more widely. Last but not least there are bioenergy and biofuels, important because they offer many of the advantages of fossil fuels, in particular being easily stored. Not surprisingly they are receiving much attention from policy makers and researchers both in the EU and North America.

Most of this renewable energy will be converted into electricity. The renewable energy resource will be geographically highly distributed, and being mostly dependent on changing weather and climate cannot be directly controlled in the way fossil fuelled generation is. Electrical power networks were designed to operate from electricity generated in a few large power stations fuelled by coal, gas or uranium, fuels readily available on the international market and to varying degrees controllable. Significantly increasing the input from renewable energy sources requires a revision of the way power systems are designed and operated in order to accommodate these variable sources better. This book is an introduction to this important topic.

The material in this book is largely based on a Master's course module taught for over ten years at the Centre for Renewable Energy Systems Technology (CREST) at Loughborough University. The course as a whole was designed to provide general technical education in all major electricity generating renewable energy sources and their integration in electrical networks. Students taking this course normally have first degrees in numerate topics ranging from Physics or Engineering to Environmental Science. The course modules are therefore designed for students who, although they may be very knowledgeable in their speciality, will only have elementary knowledge of other topics.

Likewise, this book assumes no previous knowledge in power systems engineering and guides the reader through the basic understanding of how a power system is put together and the way in which it ensures that the consumer demand is met from instant to instant. The characteristics of traditional and renewable energy (RE) resources are described with special reference to the variability of the latter and the way this impacts on their utility. These resources are available in a form that either has to be converted into electricity and/or their electrical output has to be conditioned before it can be fed into the grid. The book covers these aspects and stresses the importance of power electronic technology in the process of power conditioning. The power flows in an electricity network have to be appropriately controlled and the book addresses the way this is achieved when these new sources are integrated. The economics of renewable sources will determine their take-up by the market, and this issue is also addressed, and in some detail. Finally, an eye is cast on the future development of RE technologies and the way that power systems may evolve to accommodate them. An Appendix is available for readers who require a more mathematical coverage of the way electricity is generated, transported and distributed to consumers.

Acknowledgements

This book contains input from other CREST staff besides the main authors. In particular, Dr Murray Thomson has provided much of the power electronics material of Chapter 4 and most of the content of Chapters 5 and 6. His comments and criticisms during the initial development of the book have been invaluable. In addition, Dr Simon Watson contributed most of the material found in Chapter 7. The material on dynamic demand control in Chapter 3 was the subject of a CREST Master's dissertation by J. A. Short. Finally, thanks are due to Dr Graham Sinden who gave us permission to use several diagrams from his recent work including some unpublished ones from his doctorate thesis and to Mr David Milborrow for permitting us to use in Chapter 7 several of his tables and figures.

We are also grateful for the support of the Wiley staff in Chichester who have guided us in the process of preparing the manuscript for publication.

Last, but certainly not least, we would like to dedicate this book to our respective partners Delphine Freris and Marion Peach who have had to put up with us slaving over the text in our spare time, rather than participating more fully in family life.

Leon Freris and David Infield

1

Energy and Electricity

1.1 The World Energy Scene

1.1.1 History

Energy demand in the pre-industrial world was provided mostly by man and animal power
and to a limited extent from the burning of wood for heating, cooking and smelting of metals.
The discovery of abundant coal, and the concurrent technological advances in its use, pro-
pelled the industrial revolution. Steam engines, mechanized production and improved transportation, all fuelled directly by coal, rapidly followed. The inter-war years saw the rise of
oil exploration and use. Access to this critical fuel became a key issue during the Second
World War. Post-war industrial expansion and prosperity was increasingly driven by oil, as
was the massive growth in private car use. More recently a new phase of economic growth
has been underpinned to a great extent by natural gas.

   A substantial proportion of coal and gas production is used to generate electricity, which
has been widely available now for over a century. Electricity is a premium form of energy
due to its flexibility and ease of distribution. Demand worldwide is growing, driven by the
explosion in consumer electronics, the associated industrial activity and the widening of
access to consumers in the developing world.

1.1.2 World Energy Consumption

The present global yearly *primary energy*[1] consumption is, in round figures, about 500 EJ.[2]
This is equivalent to about 1.4×10^{17} Wh or 140 000 TWh . Dividing this figure by the number
of hours in the year gives 16 TW or 16 000 GW as the average rate of world primary power

---

[1] Primary energy is the gross energy before its transformation into other more useful forms like electricity.

[2] The unit of energy in the SI system is the joule, denoted by J. Multiples of joule are kJ, MJ, GJ, TJ (T for tera
denoting 10^{12}) and EJ (E for Exa denoting 10^{18}); the unit of power is the Watt (W) and represents the rate of work
in joules per second. Electrical energy is usually charged in watt-hours (Wh) or kWh. Joules can be converted into
Wh through division by 3600.

Renewable Energy in Power Systems   Leon Freris and David Infield
© 2008 John Wiley & Sons, Ltd

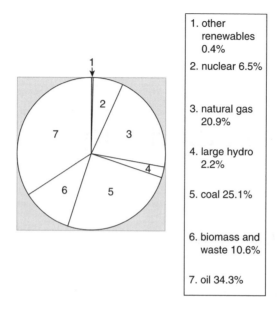

Figure 1.1   Percentage contribution to world primary energy

consumption. The pie chart in Figure 1.1 shows the percentage contribution to world primary energy from the different energy sources according to data taken from the International Energy Agency (IEA) Key World Energy Statistics, 2006.

The world demand for oil and gas is increasing significantly each year. The major part of this increase is currently taken up by India and China where industrialization and the demand for consumer products is escalating at an unprecedented pace. The world consumption in 2006 increased by more than twice Britain's total annual energy use and is the largest global yearly increase ever recorded. China alone accounted for roughly 40% of this increase. The IEA forecasts that by 2030 demand for energy will be some 60% more than it is now.

1.1.3 Finite Resources

It is extremely difficult to determine precise figures on the ultimate availability of fossil fuels. According to the major oil and gas companies, still significant new resources of oil are being developed, or remain to be discovered. A safe assessment is that there is enough oil from traditional sources to provide for the present demand for 30 years. The latest figures for global gas reserves indicate that these are approximately 50% higher than oil at some 60 years of current demand, and gas is far less explored than oil so there is probably more to be found. There are, however, unconventional hydrocarbon resources such as heavy oil and bitumen, oil shale, shale gas and coal bed methane – whose total global reserves have been assessed very roughly to be three times the size of conventional oil and gas resources. These are more expensive to extract but may become exploitable as the price of fossil fuels increases due to the steady depletion of the more easily accessible reserves. Fortunately for fossil fuel dependent economies, coal reserves are considered to be many times those of oil and gas and could last for hundreds of years. The downside of coal is its high carbon content, a topic to be discussed later.

Much debate is currently focused on when the so-called peak oil and gas might occur. This is when the oil and gas extraction rate starts to fall and occurs well before resources run out. It is important because it signals that demand will most likely not be fully met, with prices rising significantly as a consequence. Certainly the UK's North Sea reserves of oil and gas are fast declining with peak extraction having already occurred in 2003. Given the enormous investment in extraction and supply infrastructure, and the profits to be made, it would be surprising if those with vested interests did not work hard to maintain confidence in these sources.

Fuel for nuclear fission is not unlimited and several decades ago this has prompted interest in the fast breeder reactor which in effect extends the life of the fuel. However, the political dangers inherent in the fast breeder cycle, with its production of weapons grade plutonium, has limited its development to a few prototype reactors which had major operational problems and are now defunct. The lifetime of uranium reserves for conventional fission at current usage has been estimated by some as around 50 years, but such calculations are very dependent on assumptions. If an extremely high ore price is tolerable, then very low grades of uranium ore can be considered as possible reserves. The DTI cites OECD/NEA 'Red Book' figures to claim that based on 2004 generation levels, known uranium reserves (at \$130/kg) will last for around 85 years (see References [1] and [2]).

1.1.4 Energy Security and Disparity of Use

Energy security is a major concern worldwide. A large part of the world's oil is located in the Middle East and other politically unstable countries. The conflict between 'Western' and 'Islamic' cultures is at present exacerbating the anxiety over reliability of energy supply. Russia is a major producer of gas but recent events in Ukraine have made European countries aware how dependent they are on this single source. The USA is the world's largest consumer of energy and is heavily dependent on imported oil. With economic growth seen as being intrinsically linked to cheap fuel it is difficult to imagine political parties, in the USA or elsewhere, proposing policies that require voters to drastically curtail their consumption and therefore alter their lifestyles.

Another disturbing aspect is the disparity in consumption between rich and poor countries: the richest billion people on the planet consume over 50% of all energy, while the poorest billion consume around 4%. This is an added source of tension and of accusations that the developed countries are profligate in the use of energy. To excuse this high consumption on grounds of high industrial activity is simply wrong. Japan, for example, is the world's second largest economy but has a per capita energy consumption half that of the US.

1.2 The Environmental Impact of Energy Use

1.2.1 The Problem

Fossil fuels have one thing in common: they all create carbon dioxide when burnt. They are a key part of the Earth's long term carbon cycle, having been laid down in geological periods when the climate was tropical across much of the planet and atmospheric CO_2 concentrations were very high. This storing of carbon through the growth of plant matter, and its subsequent conversion to coal, oil, peat and gas, dramatically reduced atmospheric CO_2 levels and played an important role in cooling the planet to temperatures that could support advanced life forms. The concern now is that by unlocking this stored carbon climate change is being driven in the other direction, with global warming the direct result of an excessive greenhouse effect.

Ice core samples indicate that the level of carbon dioxide in the atmosphere was more or less stable at 280 parts per million (ppm) over the last few thousand years up to the onset of the industrial revolution at the beginning of the nineteenth century. Subsequently, atmospheric CO_2 levels rose, at first slowly as a result of coal burning but since the Second World War the release of CO_2 has accelerated reflecting the exploitation of a wider range of fossil fuels. Current CO_2 levels are 380 ppm and rising fast.

CO_2 is not the only pollutant created by fossil fuelled generation: combustion in air comprising 78% nitrogen by volume inevitably produces nitrogen oxides, NO, and NO_2 and N_2O, collectively known as NO_x; and any sulfur content of the fuel results in SO_x emissions. NO_x and SO_x together contribute to acid rain and as a result it is now common to reduce any SO_x emissions from fossil fuelled power stations through flue gas desulfurization. The downside of this is reduced thermodynamic efficiency and some resulting increase in CO_2 emissions.

World coal reserves are substantial, but coal is a less attractive fuel from the point of view of CO_2 emissions and also much more disruptive to extract. The cheapest coal is from opencast mines, but this process is immensely damaging to the environment. All forms of generation have some environmental impact, but these are not in general reflected in the cost of electricity; because of this, these additional environmental costs are known as *externalities*.

Externalities are consequences of activity that are not normally a part of the economic analysis; for example the cost to society of ill health or environmental damage arising from pollution caused by a specific generating plant is not directly charged to the operator, i.e. it is external to the microeconomics of the plant's operation. A number of European countries now seek to bring these externalities back into the economics of electricity generation by some kind of environmental levy or carbon tax. Carbon trading, discussed in detail in Chapter 7, is an alternative means of achieving this goal.

The nuclear cycle is of course not without externalities, although the environmental costs are highly contested, contributing as they do to the economic attractions or otherwise of nuclear power. Radioactive waste disposal, radioactive emissions and final decommissioning and disposal of radioactive reactor components are rarely fully accounted for and thus fall to an extent into the category of externalities. There are also issues concerned with environmental damage associated with uranium mining, but in this regard it is similar to coal. If nuclear power is to mitigate global emissions, it is of vital importance to assess accurately how much CO_2 will be displaced by nuclear power. This is a topic fraught with controversy. The well established 386 g CO_2/kWh contributed by gas fuelled power stations will be taken as the benchmark. The emissions for nuclear power are quoted as 11–22 by OECD, and 10–130 by ISA, University of Sydney [3]. If the upper figures are valid, the contribution of nuclear power to CO_2 mitigation may be seriously compromised. Clearly this is an issue that requires certainty.

1.2.2 The Science

The science of climate change is very well established and its primary goal is to understand the link between CO_2 and other greenhouse gas concentrations and temperature rise. Work in this area has been carried out by the *Intergovernmental Panel on Climate Change (IPCC)*, which was set up in 1988 by the World Meteorological Organisation and the United Nations Environment Programme. It involves scientists from 169 countries.

Figure 1.2 shows the changing average global temperature, from 1850 to 2005. The bold curve is the smoothed trend while the individual annual averages are shown as bars. The temperatures are shown relative to the average over 1861–1900. The earth has warmed by 0.7 °C since around 1900, bringing the global temperature to the warmest level in over 12 000 years. All ten warmest years on record have occurred since 1990 and there is considerable physical and biological evidence confirming climate change. Most climate models indicate that a doubling of greenhouse gases since the pre-industrial period is very likely to result in a rise between 2–5 °C in global mean temperatures. This increased level is likely to be reached between 2030 and 2060. If no action is taken concentrations would be more than treble pre-industrial levels by 2100, resulting in a warming of 3–10 °C according to the latest climate projections.

Although the relationship between CO_2 concentration, temperature change and undesirable climatic changes is very complex and thus hard to predict precisely, it is widely believed that the CO_2 concentrations have to be stabilized if damaging global warming is to be avoided.

The IPCC concluded in 2001 [4] that there is strong evidence that most of the warming observed over the last 50 years is *anthropogenic* in that it is attributed to human activities. This was supported by the Joint Statement of Science Academies (2005) and a report from the US Climate Change Science Programme (2006). An IPCC updated report which was published in 2007 confirmed this link with greater certainty. A summary of recent scientific research may be found in Reference [5].

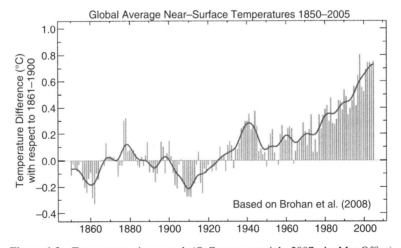

Figure 1.2   Temperature rise record. (© Crown copyright 2007, the Met Office)

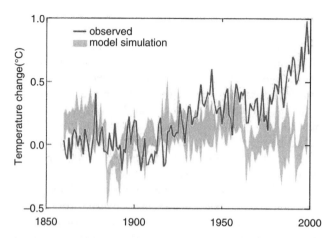

Figure 1.3 Natural factors cannot explain recent warming. (© Crown copyright 2007, the Met Office)

The basic evidence that provides confirmation that global warming is due to human-made factors was provided by a climate model developed at the Hadley Centre for Climate Prediction and Research in the UK. In Figure 1.3, the observed global temperature since the early 1900s is shown by the bold line. The climate model was driven over that period by natural factors such as output of the sun, changes in the optical depth of the atmosphere from volcanic emissions and the interactions between the atmosphere and oceans. The predictions of the model are shown by the fuzzy band. This clearly disagrees with the observations particularly since about 1970, that observed temperatures have risen by about 0.5 °C, but those simulated by natural factors have not changed at all.

If the climate model is now driven by natural factors as previously but in addition by man-made factors – change in greenhouse gas concentrations and sulfate particles – the model simulation predictions in Figure 1.4 are in much better agreement with the temperature record. Climate modelling studies by other research centres have arrived at the same broad conclusions.

1.2.3 The Kyoto Protocol

The effects of climate change are global and hence mitigation requires coordinated international effort. Signed in 1997, the Kyoto Protocol aims at reducing greenhouse gas emissions in the period 2008 to 2012 to 5.2% below those in 1990. Emissions of greenhouse gases by the US are currently 20% higher than in 1990 while the target figure in Kyoto was a cut of 7%. In the long run however it is prudent for industrialized countries to reduce their emissions by 60% by 2050 if the worst effects of climate change are to be mitigated with any confidence. This is a major challenge, to individuals, to governments and to supranational bodies. Greatest responsibility rests of course with the nations producing the largest CO_2 emissions per capita and those moving fast up the emissions table. Table 1.1 illustrates the variation in emissions per head and how this is partly driven by the income per head. Emissions from China are expected to surpass those of the US by 2025 so there is much to be done.

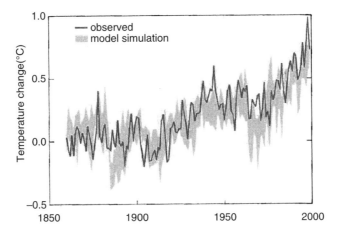

Figure 1.4 Climate warming can be simulated when man-made factors are included. (© Crown copyright 2007, the Met Office)

Table 1.1 Energy related CO_2 emissions. (Reproduced with permission from Climate Analysis Indicators Tool (CAIT) version 4.0 (or 5.0). World Resources Institute, 2007, available at http://cait.wri.org)

Country/grouping	CO_2 per head (tCO_2)	GDP per head ($)
USA	20.4	34 430
EU	9.4	23 577
UK	9.6	27 276
Japan	9.8	26 021
China	3.0	4 379
India	1.1	2 555
World	4.0	7 649

1.2.4 The Stern Report

The economics of climate change mitigation are crucial in steering an optimal policy towards a given agreed goal. These issues were addressed in the Stern Report [6].

Although the experts are almost universally convinced that climate change is taking place they are uncertain as to what exactly will be its effects. The Stern Report is unique in the sense that it examines the probabilities of reaching certain temperature thresholds at different stabilization levels. These probabilities have only been established recently and provide the basis for the economics of the analysis of the risks and costs involved in taking a range of actions towards reducing the greenhouse emissions.

The key message of the Stern Report is summarized below:

- The lags in the climate change process must be recognized. What is going to happen to the climate over the next 20–30 years is already determined and irreversible. Actions over the next 20–30 years will affect what happens in the decades to come.

- Climate change threatens the basic elements of life, i.e. access to water, food, health and the use of land and the environment.
- There is still time to avoid the worst impacts of climate change if action is taken now.
- Stabilization at 550 ppm of all greenhouse gases is recommended, but this would involve strong action.
- The costs of stabilizing the climate are significant (1% of global GDP) but manageable. Delay would be dangerous and much more expensive, perhaps as costly as 20% of global GDP.
- Action demands an international response.

The key actions should include:

- Increase in efficiency of energy use.
- Strict emissions trading rules to support the transition to low carbon development paths.
- Extensive use of renewable and other low carbon technologies.
- Technology cooperation and fivefold increased in low carbon technologies R&D.
- Reduction in deforestation.

The major focus of the Stern Report is the economics of climate stabilization. Figure 1.5 shows estimates of costs of low carbon technologies in 2015, 2025 and 2050 that may be used to constrain CO_2 emissions. The costs are expressed as a central estimate, with a range, and as a percentage of the fossil fuel alternative in the appropriate year. Due to learning effects the costs fall over time. The ranges reflect judgements about the probability distribution of unit costs and the variability of fossil fuel prices. The 0% line indicates that the costs are the same as the corresponding fossil fuel option. As expected, the uncertainties are large even for short term predictions. Onshore wind is shown to be particularly attractive with photovoltaic (PV) cells becoming very attractive beyond 2025.

On the basis of the costs of the low carbon technologies and assumptions on possible rates of uptake over time, the Stern Report estimates the distributions of emissions savings by technology for 2025 and 2050 for the desirable climate stabilization at 550 ppm. These estimates are shown in Figure 1.6. Energy efficiency and carbon capture and storage (CCS) play a major role in this scenario and will be discussed later in this chapter. Contributions from wind, solar, biofuels, hydro and distributed combined heat and power (dCHP) through electricity generation provide the remaining savings; and these are the technologies to be addressed in later chapters of this book.

1.2.5 Efficient Energy Use

Figure 1.6 stresses that efficiency measures are projected to make the largest contribution in climate change mitigation. It is therefore a surprise that the important topic of rational and efficient use of energy is rarely pursued vigorously in national or supranational plans in spite of the fact that study after study has shown that this route provides the most cost effective way to meet sustainability goals.

In most countries, regulations and financial incentives are now in place to encourage energy efficiency but their effect is modest and national energy consumption figures continue to

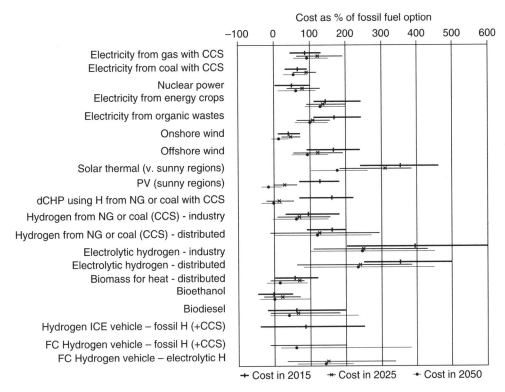

Figure 1.5 Unit costs of energy from low carbon technologies: CCS stands for carbon capture and storage, dCHP stands for distributed combined heat and power. (Reproduced from Stern review website, copyright Cambridge University Press)

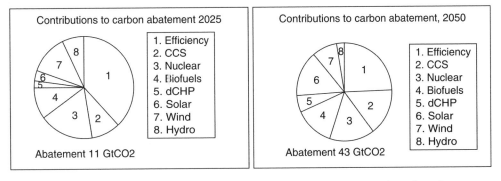

Figure 1.6 The distribution of emission savings by technology. (Reproduced from Stern Report, copyright Cambridge University Press)

rise year on year. Energy efficiency must be the linchpin of any future energy strategy because [7]:

- Using energy as efficiently as possible is the most cost effective way to manage energy demand, and thus to address carbon emissions. Saving energy is cheaper than making it.

- By reducing demand on gas and electricity distribution networks, energy efficiency will improve the security and resilience of these networks and reduce dependence on imported fuels.
- By reducing energy bills, energy efficiency will help businesses to be more productive and competitive.
- Improving the energy standards of homes has an important role in reducing spending on fuel by those in fuel poverty.

Increasing energy end use efficiency is unattractive for energy companies driven by commercial imperatives to increase sales and profits. It thus falls to governments to implement policies that change these drivers. Regulations can be put in place for example that require utilities to encourage customers to use electricity efficiently. A more revolutionary approach envisages the utility being transformed into a supplier of energy services, owning appliances in people's homes and thus being motivated to maximize the efficiency of these appliances. Whatever approach is finally adopted, the importance of reducing energy consumption should be the cornerstone of any CO_2 mitigation programme.

1.2.6 The Electricity Sector

Figure 1.7 shows the percentage of fuels used in the generation of electricity. Fossil fuels account for 64% of the fuels used in this sector with coal being the dominant source at approximately 40% and contributing nearly three quarters of CO_2 emissions. At present, large hydropower plants account for the major part of the renewables sector. Under half of the electricity produced is used in buildings, about a third in industry, under one-tenth in energy production (e.g. refineries) and less than one-tenth in transmission and distribution.

The world annual generation of electricity is in the region of 18 000 TWh representing an average rate of consumption of around 2000 GW. This electrical energy is generated in a very large collection of power stations driven mostly by fossil fuels. The electricity sector is the fastest growing source of emissions and estimated to increase fourfold between now and 2050. According to Stern this sector would need to be at least 60% decarbonized by 2050

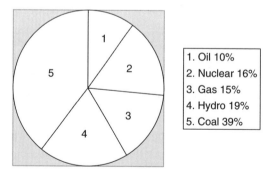

Figure 1.7   Contributions in the generation of electricity. (Data from Boyle, G., *Renewable Energy*, Oxford University Press, 2004)

for atmospheric concentration to stabilize at 550 ppm, thereby reducing the risk of catastrophic climate change.

1.2.7 Possible Solutions and Sustainability

Fundamental choices will have to be made in the years ahead. Societies are presently dependent on high and growing fossil fuel consumption. The possibility of weaning people from this dependence over a short timescale is completely unrealistic. The general shift in fossil fuels over the last two decades for both electricity generation and heating has been towards increased use of gas in place of coal, and to a lesser extent oil. This has helped to limit the growth in CO_2 emissions as gas combustion releases less CO_2 per unit of energy than coal. Political events, however, have generated anxiety in the EU and elsewhere in relation to increased dependence on this particular fuel.

A possible alternative path is to revert to dependence on coal. This resource is abundantly available in many developed counties including the US, Australia, many EU countries, Canada, Russia and in developing countries such as China, South Africa and Turkey. Recent developments in CO_2 capture or 'sequestration' for fossil fuels, discussed later in this chapter, give some hope that this source may be made more acceptable environmentally.

A number of potentially carbon neutral sources exist: these include nuclear fission (and possibly fusion in the far future) and all sources that derive directly or indirectly from the sun, namely biomass, wind, solar (thermal and photovoltaic), hydroelectric and marine. Geothermal and tidal energy are also carbon neutral and often regarded as renewable on the grounds that the sources are so huge as to be virtually inexhaustible. In Chapter 2, the characteristics of all these conventional and emerging technologies are discussed in some detail.

Finally, but no less important, other approaches essential in the move towards sustainability are a reduction in energy needs and improvements in the efficiency of energy use. The latter includes more efficient electricity generation. Although not the main focus of this book, these topics are briefly discussed in this chapter.

The planet's reserves of fossil fuels and minerals are of course finite, and thus the exploitation of coal, oil, gas and uranium are not sustainable in the longer term. Fortunately, renewable energy (RE), being derived from naturally occurring energy flows, is inexhaustible and has no long term detrimental effect on the environment. As such it will in time become the basis of the energy supply system, and probably the sole means by which electricity is generated.

1.3 Generating Electricity

1.3.1 Conversion from other Energy Forms – the Importance of Efficiency

Figure 1.8 shows the ways in which various types of energy can be converted into electricity. At present, the path generating the bulk of electricity worldwide is shown by the bold lines that lead through combustion from chemical to thermal, from thermal to mechanical and finally to electrical power conversion. The bottleneck of this path is the limited thermodynamic efficiency determined by the Carnot cycle. Older thermal generating stations have

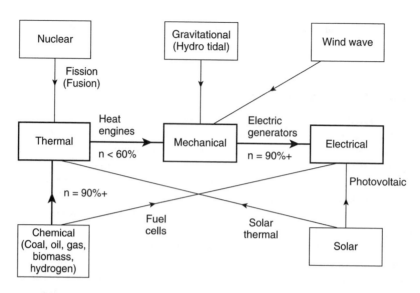

Figure 1.8 Conversion from a variety of energy forms into electricity

efficiencies between 35 and 40% although in the last two decades conversion has been substantially improved to over 50% through the development of combined cycle gas turbines (CCGTs) a technology discussed in Chapter 2. It follows that when coal, oil or gas is used only 35–50% of the primary energy is successfully converted, the remaining being discharged into the environment in the form of waste heat.

One way of getting around the Carnot limit is simply to make use of the waste heat. This is the principle of combined heat and power (CHP), used extensively in Scandinavia and of growing importance elsewhere. In such schemes, the waste energy from the thermal generation of electricity is distributed through heat mains to local industry and/or housing. This requires substantial infrastructure and is therefore only viable if the power station is reasonably close to the heat users. An alternative arrangement made possible by recent developments is to transport the fuel (mainly gas) to the consumer using the existing supply infrastructure and install the CHP system at the consumer's premises. Such systems are known as *micro-CHP* and are discussed in Chapter 8.

Direct paths that bypass the Carnot bottleneck are also available. The leading example of this approach is the fuel cell, which now borders on commercial viability in a number of forms: solid oxide, molten carbonate, and proton exchange membrane (PEM) to name the main ones. Another direct path is through photovoltaics, a technology that perhaps is the most promising in the near to far future.

The conversion efficiencies of the various routes indicated in Figure 1.8 are dealt with in greater detail in Chapter 2.

1.3.2 The Nuclear Path

The topic of electricity generation from nuclear power elicits strong emotions from supporters and critics. Although nuclear power supplies only the equivalent of 5.7% of the world primary energy at the time of writing this book, some believe this should be expanded massively. They argue that it is an attractive source of electricity, having very low carbon emissions.

After the Three Mile Island and the Chernobyl accidents there was a period of nearly ten years during which almost no new nuclear capacity was constructed. However, the recent concerns regarding fossil fuel security have prompted a number of countries to consider new building programmes. China and India are planning to build several tens of reactors each and the USA is posed to do the same. In contrast within Europe, only Finland has embarked on the construction of a new nuclear plant while, Sweden, Switzerland and Germany all have moratoriums in place leading to a phasing out of nuclear power. France on the other hand, remains committed to nuclear power which contributes about 80% of its present electricity needs.

In the UK the 2003 government White Paper was critical of the nuclear option, but by 2006, with concerns about a possible energy gap, the government's position had changed. It is now supportive in principal of a new nuclear programme. A key concern is that major investments in nuclear will deprive renewable energy sources of the finance they need to expand. Reflecting its importance, the debate over nuclear power is extensive and there is voluminous literature. Reference [8] provides a good entry point for those interested.

1.3.3 Carbon Capture and Storage

Figure 1.6 indicates, that according to Stern, by 2025 and 2050 about 20 and 40% respectively of carbon abatement is expected to be provided by the emerging technology of carbon capture and storage (CCS) provided that the technoeconomic and environmental issues can be satis-factorily dealt with. CCS has the significant advantage of reconciling the necessary use of fossil fuels in the medium term with the necessity of serious cuts in CO_2 emissions.

A large scale demonstration project was being planned in northeast Scotland, a joint venture by BP, Shell and ConocoPhilips [9]. Unfortunately, this project was recently aban-doned, but the technology is being vigorously pursued in other projects worldwide. Figure 1.9 illustrates the geological storage options for CO_2.

With gas and oil prices likely to rise significantly, extracting CO_2-free energy from coal is also attracting substantial attention. Such technologies take a number of forms, but the so-called integrated gasification combined cycle (IGCC) process is in the forefront. This involves the production of a synthetic gas (syngas) obtained from coal through gasification. Syngas is composed mainly of hydrogen and carbon monoxide and is the fuel source to a high efficiency plant operating in the combined cycle mode. Buggenum, a 253 MW plant in the Netherlands operates on this principle and is the cleanest coal based plant in Europe. To date, no IGCC plant involving carbon capture has been built although this principle is to be used as a basis for a zero carbon emission 275 MW plant funded by the US Department of Energy which is being built and should be up and running in 2013.

1.3.4 Renewables

Figure 1.10 provides an overview of the earth's main energy paths that can be tapped to generate sustainable electricity. The main source of easily accessible renewable energy

Geological Storage Options for CO_2

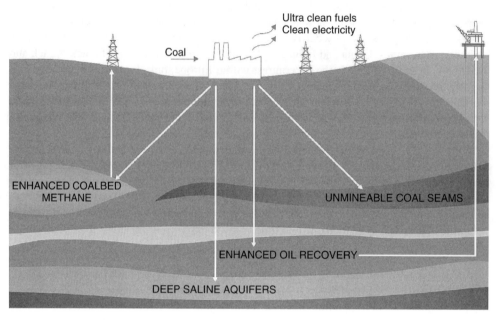

Figure 1.9   Geological storage options for CO_2. (Source: World Coal Institute)

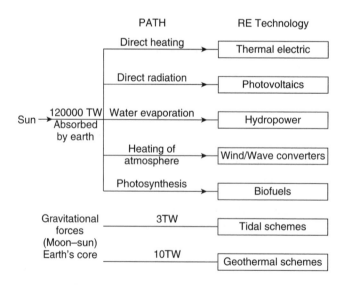

Figure 1.10   Renewable energy flow paths

is the sun. On average the rate of solar radiation intercepted by the earth's surface is about 8000 times as large as the average rate of world primary energy consumption. With the present world population this amounts to a staggering average power of 20 MW per person.

The figure shows that this energy flux can be accessed directly using solar thermal or photovoltaic technology, or indirectly in the form of wind, wave, hydro and biofuels. Two other energy sources are often regarded as renewable in view of their sustainable nature: energy in the tides caused by the gravitational fields of the moon and the sun which can be tapped using tidal barrages or tidal stream technology; and geothermal energy from the earth's core accessible in some locations through hot springs, geezers or boreholes. The available average power from these resources is a small fraction of that available from the sun.

A substantial proportion of the incident radiation is reflected back to space. Over the last several millennia and up to the onset of the industrial revolution, energy inputs and outputs have been in equilibrium at a global temperature level suitable for the development of the earth's biosphere. Exploiting the incident energy through the application of renewable energy technology *does not* disturb this balance. Intercepted natural energy flows, for example converted to electricity and then converted again by consumers into mechanical, chemical or light energy, all eventually degrade into heat.

Most renewable energy forms are readily converted to electricity. Solar energy, geothermal energy and biomass can also be used to supply heat. Renewable energy can in principal provide all the energy services available from conventional energy sources: heating, cooling, electricity and, albeit with some difficulty and cost, transport fuels. It has the additional advantage that being a naturally distributed resource, it can also provide energy to remote areas without the need for extensive energy transport systems. It is worth noting that it is not always necessary to convert the renewable energy into electricity. Solar water heating and wind-powered water pumping are fine examples of systems that can work very well without involving electricity at all. However, the major contribution that renewable energy will be increasingly making in supplying people's needs will be in electrical form.

Renewable energy is currently experiencing dramatic growth. Wind power and solar PV are leading the growth with global companies such as GE and Siemens entering the wind energy market, and BP and Shell playing a major role alongside Japanese companies like Sharpe and Sanyo in PV. In China five of the largest electrical aerospace and power generation equipment companies have recently begun to develop wind turbine technology. Most large oil companies have expanded their research and development in ethanol and biodiesel production from biomass. The fastest growing RE technology is currently grid connected PVs with 40% annual year on year growth, but the RE technology that has made the largest contribution to date (excluding conventional hydro) is wind power with over 60 GW installed in EU countries and 95 GW worldwide by the end of 2007.

At least 48 countries have national targets for RE supply including all 25 EU countries. Figure 1.11 shows the intended increase in contribution of the EU countries from 2002 to 2010. The EU has Europe-wide targets of 21% electricity and 12% of total energy by 2010. Table 1.2 shows the intentions of the EU over a wider period, with expected contributions in TWh per annum from various RE technologies.

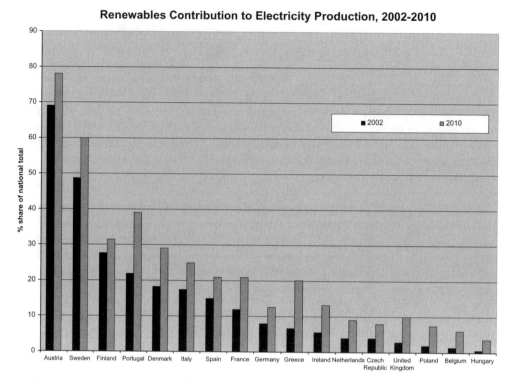

Figure 1.11 RE contribution to European electricity production. (Source: Oxford Intelligence)

Table 1.2 Contribution of renewables to electricity generation (1995–2020). (Source: Eurostat http://epp.eurostat.cec.eu.int)

	1995 Eurostat	2000 Eurostat	2010 Projections	2020 Projections
Wind (TWh)	4	22.4	168	444
Photovoltaics (TWh)	0.03	0.1	3.6	42
Biomass (TWh)	22.5	39.2	141	282
Hydro (TWh)	290	322	355	384
Geothermal (TWh)	3.5	4.8	7.0	14
Total RES in EU 15	320	388	675	1166
Total electricity[a] (TWh)	2308	2574	3027	3450
Share of RES (%)	13.9	15.1	22.3	33.8

[a] EU trends to 2030.

1.4 The Electrical Power System

1.4.1 Structure of the Electrical Power System

Electricity is widely used because it is a supremely flexible form of energy. It can be readily and efficiently transported and is easily converted to other forms of energy. Mechanical energy can be provided by very efficient motors, light energy by increasingly efficient light

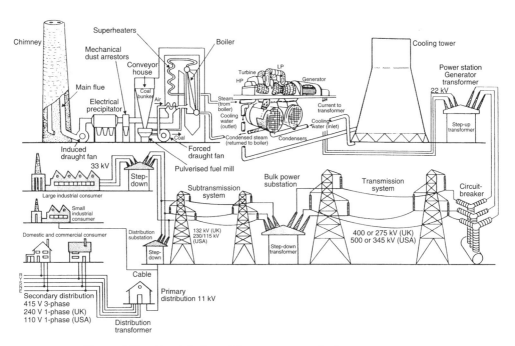

Figure 1.12 Pictorial view of the components of a large power system

fittings, heat energy by 100% efficient resistive elements, and power supply to electronic and IT (information technology) hardware through very efficient power conditioning units.

Figure 1.12 shows a diagrammatic layout of a typical electrical power system from the point of generation to the point of consumption. The figure depicts a coal fired power station as this represents the majority of world stations. The energy conversion chain follows the chemical → thermal → mechanical → electrical path depicted in Figure 1.8. Coal is pulverized and fed into a boiler where it is mixed with forced air and combusted. The boiler is a complex structure consisting of many stages of energy extraction from the combusted fuel. The flue gases are guided through equipment that removes solid particles and sulfur (desulfurization is not shown in the figure) before being released into the atmosphere. The highly purified water in the boiler is converted into superheated steam which is passed through several turbine stages on the shaft of a turbogenerator. The low pressure low temperature steam from the outlet of the turbine is condensed into the purified water which in this closed system is pumped back into the boiler. The condensing process unfortunately needs a substantial amount of external cooling water. In the figure, this water is provided from a pond at the bottom of a cooling tower. The hot water from the condenser is sprayed at the top of the tower and transfers its heat to the air that passes up the venturi shaped tower. The lost water must be made up from some external source such as a local river.

The energy generated at the power stations is transmitted to consumers by overhead transmission lines and underground cables that possess ohmic resistance. The energy loss due to the unavoidable resistive heating of a line or cable is proportional to the square of the current I it carries. Additionally, for a given power transfer, which is proportional to VI, the current is inversely proportional to the voltage (other things being equal). Thus, the loss decreases with the square of the voltage *V*. The downside of operating at higher voltages is that costs of insulation and other power system equipment increase substantially. Thus, for bulk transmission of power over long distances, higher voltages are most economic whereas, for local distribution of modest power to numerous connection points, lower voltages are most economic. The economics also dictate that it is worthwhile to have several intermediate voltages. This multiple voltage arrangement results in network transmission losses confined to within 5–10% of the throughput power.

The bulk of global electricity is generated in large (>500 MW) power stations at around 20 kV. This is then stepped up by transformers to an extra high voltage (EHV) level such as 400 kV and carried by the transmission system to the bulk supply points, where it is stepped down to a high voltage (HV) level of around 100 kV. Some very large industrial consumers are connected at this level but most power is transformed down again to medium voltage (MV) levels such as 30 kV, then to 10 kV and finally to the low voltage (LV) level of 400 V, also referred to as the distribution system, which provides 230 V, when the connection is single-phase. In the USA and a few other countries, the LV level is 200 V three-phase, 115 V single-phase. The voltages used vary from country to country but the power system structure follows closely the layout of Figure 1.12.

In Figure 1.12, a circuit-breaker is shown after the generator step-up transformer. This is a component part of an extensive protection network which permeates all levels of the power system. Faults on the network may result in low resistance paths that cause excessive currents capable of damaging equipment. The protection devices, circuit breakers at high voltage levels and fuses at domestic distribution level, operate to isolate the faulty part of the network and prevent equipment damage. The effect on the protection system of introducing renewable energy sources will be discussed in Chapter 6.

1.4.2 Integrating Renewables into Power Systems

The term *grid* is often used loosely to describe the totality of the network. In particular, *grid connected* means connected to any part of the network The term *national grid* usually means the EHV transmission network.

Integration specifically means the physical connection of the generator to the network with due regard to the secure and safe operation of the system and the control of the generator so that the energy resource is exploited optimally. The proper integration of any electrical generator into an electrical power system requires knowledge of the well-established principles of electrical engineering. The integration of generators powered from renewable energy sources is fundamentally similar to that of fossil fuelled powered generators and is based on the same principles, but, renewable energy sources are often variable and geographically dispersed.

A renewable energy generator may be described either as *standalone* or *grid-connected*. In a standalone system a renewable energy generator (with or without other back-up generators or storage) supplies the greater part of the demand. In a grid-connected system, the renewable energy generator feeds power to a large interconnected grid, also fed by a variety of other generators. The crucial distinction here is that the power injected by the renewable energy generator is only a small fraction of that generated by the totality of generators on the grid. The distinction between standalone and grid-connected generators is a useful one but is not always clear-cut. Sometimes confusion arises when the word grid is used to refer to a relatively small standalone electrical network. This is not necessarily wrong (though it may indicate delusions of grandeur!) but it should always be clear as to the extent of the grid being referred to.

The point on the network to which a renewable energy generator is connected is referred to, for reasons to be explained later, as the *point of common coupling* (PCC).

1.4.3 Distributed Generation

Power systems have developed over the years to supply a varying demand from centralized generation sourced from fossil and nuclear fuels. Unless nuclear fusion proves successful, which will not be known for over 50 years, there is universal agreement that by the end of this century the majority of our electrical energy will be supplied from RE sources.

Generators powered from renewable energy sources (except large scale hydro and large offshore and onshore wind farms) are typically much smaller than the fossil fuelled and nuclear powered generators that dominate today's large power systems. Small generators cannot be connected to the transmission system because of the cost of high voltage transformers and switchgear. Also, the transmission system is often a long way away as the geographical location of the generator is constrained by the geographical availability of the resource. Small generators must therefore be connected to the distribution network. Such generation is known as *distributed* or *dispersed generation*. It is also known as *embedded generation* as it is embedded in the distribution network.

In traditional power systems power invariably flows from the large centralized power stations connected to the EHV network down through the HV and LV systems to be distributed to consumers. In power systems with distributed generation power may travel from point to point within the distribution system. This unusual flow pattern has some serious implications in the effective operation and protection of the distribution network. Distributed generation will be discussed in Chapter 6.

It may be concluded that present power systems will gradually have to evolve and adapt so that, in the far future, a managed demand will be supplied from distributed, mostly variable, RE generation. This transformation will be aided by the liberal use of power electronic interfaces capable of maximizing the effectiveness of RE sources, controlling power flows and ensuring reliability of supply. Some of these issues are discussed in the last chapter.

1.4.4 RE Penetration

The proportion of electrical energy or power being supplied from renewable sources is generally referred to as the *penetration*. It is usually expressed as a percentage. When fuel or CO_2-emission savings are being considered, it is useful to consider the *average penetration*:

$$\text{Average penetration} = \frac{\text{Energy from renewable energy powered generators (kW h)}}{\text{Total energy delivered to loads (kW h)}}$$

In this case, the energy (kW h) is measured over a long period of time, perhaps a year. At first sight, it might seem more natural to express the denominator (total energy delivered to loads) as: total energy from all the generators (including fossil fuelled generators). However, in standalone systems, there may be *dump loads* (loads where energy is dumped as heat) to consider and, in grid-connected systems, there is often interest in the penetration in a given geographical area, in which case it may be termed the *local penetration*.

For other purposes, including system control, it is necessary to consider the instantaneous penetration:

$$\text{Instantaneous Penetration} = \frac{\text{Power from renewable energy powered generators (kW)}}{\text{Total power delivered to loads (kW)}}$$

Since the electrical output from some generators operating from renewable energy sources is variable, the maximum instantaneous penetration will normally be much higher than the average penetration.

References

[1] Barnaby, F. and Kemp, J. 'Too hot to handle? The future of civil nuclear power', Briefing paper, July 2007, Oxford Research Group.
[2] Cited in 'The future of nuclear power', DTI/Pub 8519/4K/05/07.NP, p.156; www.dti.gov.uk/files/file39197. pdf.
[3] See www.pmc.gov.au/umpner/docs/commissioned/ISA_report.pdf.
[4] IPCC, *Third Assessment Report of the IPCC*, Cambridge University Press, 2001.
[5] Schellnhuber, H.J. (ed). *Avoiding Dangerous Climate Change*, Cambridge University Press, 2006
[6] *Stern Review Report on the Economics of Climate Change*, Cambridge University Press, 2006. Following criticisms of the report a rebuttal was published by Stern and his team and can be found in http://www.hm-treasury. gov.uk/media/5E1/FB/stern_reply_worldeconomics.pdf.
[7] 'Plan for action on energy efficiency', April 2004, UK government follow-up to the 'Energy White Paper'.
[8] 'The role of nuclear power in a low carbon economy', Sustainable Development Commission, UK, March 2006.
[9] Gordon, S. 'Carbon cure', *IEE Power Engineering*, December 2005/January 2006.

2

Features of Conventional and Renewable Generation

2.1 Introduction

The purpose of this chapter is to describe the essential features of the different electricity generation plant, so that the role of the renewable sources and how they might be integrated into the electricity supply system can be better understood. The main features of conventional sources are dealt with in one section and in less detail. Such sources supply the overwhelming proportion of energy in power systems in which renewables are now being integrated but are themselves not the focus of this book. Such plant will remain for a long time to come and in the transition to a sustainable supply system they will need to complement the growing proportion generated from renewable sources. It is therefore essential that the reader has a basic understanding of conventional sources and their characteristics. Each of the renewable energy sources is treated in greater length and separately to reflect their rather diverse characteristics.

First the rather obvious fact should be remembered that due to conservation of energy the time varying power demand of the consumers should be continuously matched by the generated power. In this situation, generation is said to be *load following*. How this is done and what happens when exact balance is not achieved is the topic of the next chapter. Conventional sources, taken here to include both fossil fuelled and nuclear generation, in general produce a given output when required to by the operator and so can be *dispatched*,[1] whereas most renewable sources generate according to the time varying strength of the renewable energy source. These time variations are particularly noticeable in the case of wind, solar, wave and tidal power, although they are different in the extent to which the variations can be predicted, a factor that will turn out to be significant for efficient operation of the power system. Conventional generation in itself is neither instantly dispatchable nor completely flexible; it takes time to prepare large thermal plant for full load output and such plant are

---

[1] Dispatching is the planned allocation of plant to meet expected future loads on the system.

Renewable Energy in Power Systems   Leon Freris and David Infield
© 2008 John Wiley & Sons, Ltd

not permitted to operate at too low a load. Nuclear plant in particular offers very limited capability to follow changes in load.

The remainder of this chapter explores in detail the characteristics of the different sources of generation with specific regard to their contribution to meeting the demands of a large power system. All the renewables are dealt with except geothermal power, which is very geographically limited, and where it does exist can be treated very much like fossil fuelled thermal power. It is after all simply a way of raising steam using large naturally occurring thermal gradients.

2.2 Conventional Sources: Coal, Gas and Nuclear

The diverse characteristics of generators operating from different fossil fuel sources define their function in an integrated power system. Nuclear powered stations, dealt with in more detail at the end of the section, are generally inflexible and designed to run at constant power. This mode of operation is known as *base load* since it contributes to that fraction of the load that can be regarded as always present.

Conventional coal plant efficiency, as reviewed in Chapter 1, is in the range 30–40% and depends significantly on how it is operated.[2] The higher efficiencies would correspond to the best plant in base load operation; increased *cycling* [3] and low load operation of the plant significantly lowers operational efficiency. Peak system loads generally occur for only short periods of time and it turns out that such loads are best met by plant of low capital cost;[4] the high operational costs associated with typically low efficiency generation of this peaking plant is acceptable since total operation time tends to be limited.

Thermal conversion has been substantially improved over the last two decades through the development of combined cycle gas turbines (CCGTs). In this arrangement the gas, usually natural gas, is burnt at temperatures of around 1000 °C to drive a turbine and the exhaust gases are subsequently used to raise steam for a traditional steam turbine. Both turbines drive separate generators feeding power to the grid. By increasing the temperature at which the fuel is burnt efficiencies of around 50% can currently be achieved, with plant efficiencies towards 60% being projected for the next generation. Such plant are currently highly controllable, but because of their high efficiency they are constrained by fuel supply contracts that dictate running on high load. Most gas powered stations are therefore run at constant power and thus also contribute to the base load, although it is worth noting that recent rises in gas prices within Europe have recently changed the situation to one where it is, at least for the time being, economically preferable to run coal fired stations at base load in place of CCGT plant.

---

[2]Most conventional modern coal plants burn pulverized coal; future plant using gasified coal could reach a considerably higher overall efficiency.

[3]Plant cycling is the process by which a plant is run up from cold to meet load and subsequently run down when not required. During the transitional periods plant is generally operating at far from optimal conditions and energy is required to prepare the plant for generation. All this contributes to poor overall efficiency.

[4]Open cycle gas turbines provide cheap peaking capacity but are relatively expensive to operate. The impact of plant costs on their role in the system is explored in more detail in Chapter 7.

The design of generation plant does affect its time response and the extent that it can contribute to the regulation demands of the power system. Load following capabilities reflect in the main the thermal mass of the central plant elements. Thus coal stations with large boilers can take many hours to reach full output from cold. In contrast, gas turbines can reach rated power in minutes and combined cycle plants lie somewhere between coal fired and direct gas fired plant. Innovations in plant design aimed at improving conversion efficiency can have the effect of reducing flexibility. This important issue is discussed in detail in Reference [1].

Although the term nuclear power formally covers both fission and fusion, fusion power is still decades away and may never be technologically viable or cost effective, and it certainly cannot be regarded as conventional. Nevertheless, considerable international research has been and continues to be directed towards this technology, in particular through the recently agreed ITER consortium. In contrast fission based reactors have been generating a significant proportion of electricity worldwide. It remains a controversial technology due to the dangers of radiation and the challenge of radioactive waste disposal. A number of different technologies have been developed including boiling water reactors and gas cooled reactors.

Such plant sustain a fission chain reaction within a controlled environment. Fission takes place in the reactor core which is normally contained within a pressure vessel safety shield. Moderators, usually graphite or water, slow down the neutrons and help regulate the chain reaction. Control rods are inserted into the core to regulate the nuclear reaction; these are made of a material such as boron which absorbs neutrons. The core also comprises fuel rods of fissile material. A coolant, normally water or gas, passes through the reactor. It then passes to the boiler where steam is raised. Electrical generation is provided by steam turbines and in this regard nuclear power stations are similar to large coal fired stations; it is simply the source of heat for raising the steam that is different.

In general nuclear reactors are regarded as inflexible and they normally operate on base load. The fact that capital and installation costs far outweigh fuel and operational costs is further motivation for operating such plant as continuously as possible. Load factors are around 80% but would almost certainly fall if they were used to any extent to load follow, as advocates of nuclear power in France have claimed is possible. The activity of load following is sometimes confused with the steady reduction of nuclear power station output over some weeks, which is possible, indeed advisable, if the station is being run down prior to maintenance.

2.3 Hydroelectric Power

As mentioned in Chapter 1, hydroelectric generation is an indirect form of solar energy. Incident solar radiation evaporates water from the sea, and to a lesser extent from land areas, and the warmed water vapour rises; as it ascends it expands and cools, eventually condensing in the form of clouds. Some of the resulting rain falls on high ground. This water has thus gained potential energy as a result of solar input. Hydro power is the result of extracting some of this energy as the water flows back towards the sea. Large scale hydro makes use of large reservoirs, usually created by damming rivers. Water is allowed to flow out of the reservoir in a controlled manner, turning turbines that drive electrical generators as it does so.

The storage of water in the reservoir allows generation to be timed to meet the demands of the power system. Energy storage capacity[5] is limited so the aim is usually to generate at times of high load and so maximize the income generated. Since water availability is limited and seasonal, complex algorithms making use of rainfall prediction are used to optimize hydro operation. Some countries, for example Norway and Switzerland, have bountiful hydro resources and as a result there are times when electricity is so cheap it is almost given away. This has encouraged the development of industries that require abundant cheap electricity. Energy intensive processes such as aluminium smelting and silicon production are often located in countries with plentiful hydro resources for exactly this reason.

Dams are attractive because they can provide a large *head* (equal to the fall in height of the water) but whether they can be built depends on the local topography. Rivers are the natural way in which water loses potential energy and it is possible to extract a proportion of this by means of so called low head schemes which are usually small scale. It is also possible to place turbines in a river flow directly with no dams or penstocks,[6] extracting only a very small amount of the predominantly kinetic energy as the river flows by. Such turbines operate at effectively zero head and the installations are known as run of river schemes.

2.3.1 Large Hydro

Large scale hydro is a well developed and widely used form of generation. Depending on estimates, between 20 and 25% of the world's large scale hydro potential has already been developed, although the resource reasonably located geographically in relation to electricity demand has according to many commentators already been largely exploited. Hydroelectric stations currently contribute about 20% of world electricity generation. Large scale hydro power is operationally the most desirable of all renewable energy sources with respect to availability and flexibility of supply.

As explained above, water can be stored in reservoirs and used when required, either continuously if the reservoir is large or when most required by the demand for electricity. The advantage of this storage arrangement is compounded by the natural capability of hydro plant to respond within minutes to demand increase or decrease. Such plant is therefore invaluable as a means of flexible generation to follow both predicted and unexpected changes in consumer demand. This feature is so valuable that in countries such as the UK where the topography is unsuitable for the installation of large conventional hydro, *pumped storage* schemes driven by the bidirectional transfer of water between two reservoirs have been developed.

The downside of large hydro schemes is that they involve substantial upfront capital investments with profits accrued over long periods in the future. Their development can also be environmentally undesirable because of the flooding of large areas and the displacement of populations, such as has occurred with the construction of the Three Gorges scheme in China. However, this established renewable energy technology is unlikely to provide substantially increased contributions in the future since many of the attractive sites have already been developed; the areas with remaining potential lie mainly in the former Soviet Union and the

---

[5] Capacity for a store is the energy that can be stored (i.e. MW h), rather than the rating.

[6] A penstock is a sluice or gate used to control the water together with the pipe to take the water to the turbine.

developing countries [2]. The optimal way of integrating this technology in power systems has been covered in many books and scientific papers published over the last century and will not be dealt with further here. Reference [3] presents an up to date approach to hydro scheduling.

2.3.2 Small Hydro

A small-scale hydro is commonly defined as being smaller than 5 MW. At even lower powers (<100 kW), the so-called microhydro is subdivided into dammed and run of the river schemes in which there is no storage; a further distinction is into high and low head. It is estimated that the world potential of small/microhydro is around 500 GW of which roughly only one-fifth has been exploited to date. A particular attraction of small/microhydro is that the resource is often located in remote rural upland areas unserved by a conventional electricity supply, but where the size of communities and their energy requirements are consistent with the available supply.

The flow in a given river will vary greatly throughout the year, generally having, in the Northern hemisphere, high values during the winter months and low values during the summer months. In more tropical climates the flow is likely to relate to monsoon conditions. These seasonal constraints are important. For example, a correctly sized small scale hydro-power scheme in the UK will not be expected to have sufficient water to run continuously throughout the summer months. For this reason most small hydro systems are grid connected.

Figure 2.1 shows the mean daily flow for the river Barle in Somerset for 1980 from January through to December. The following observations can be made:

- For two months in the summer the flow dropped to very low values, less than 1 m³/s.
- The river flooded (over 20 m³/s for this river) on several occasions.
- The river level tends to rise fairly quickly but this level reduces gradually; i.e. there is a sharp 'leading edge' followed by a slow decay. This is a general characteristic of rivers.

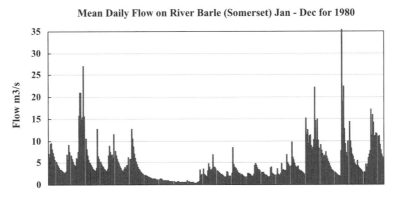

Mean Daily Flow on River Barle (Somerset) Jan - Dec for 1980

Figure 2.1 The mean daily flow on the River Barle at Dulverton Somerset in 1980. (Provided by Robin Cotton for CREST MSc notes, Loughborough University, 1998)

Small scale hydro schemes show little variability in output from minute to minute, but can change substantially over hourly or daily cycles due to sudden rainfall. As already mentioned, the output is also highly sensitive to the time of year.

If a large number of small scale hydro schemes are connected to an integrated electricity network their influence on the minute to minute operation of the power system would be negligible due to the statistical smoothing effect due to the aggregation of uncorrelated short term variations.

Turbine Designs

All hydo systems exploit the potential energy of the water. Engineers use the term head, H, for the height through which the water is allowed to fall; often in practice this becomes the net or effective head, reflecting frictional losses. Potential energy is simply mgH where g is the acceleration due to gravity; for a volume V of water density ρ this is $V\rho gH$. The power in kilowatts associated with a flow rate $Q\,\mathrm{m}^3/\mathrm{s}$ falling through an effective head of H metres is thus given by:

$$P = Q\rho gH$$

Two different approaches to turbine design exist: impulse turbines like the Pelton wheel extract kinetic energy from the flow through impact on cups mounted on the turbine wheel, while reaction turbines like Francis and Kaplan designs, which in contrast to impulse turbines, operate submerged in water. Impulse turbines are suited to high pressure/head and can have efficiencies around 90%, whereas reaction turbines run faster and are suited to lower heads and have higher efficiencies. Turbine efficiency is conventionally plotted, as in Figure 2.2, as a function of the specific speed $N_S = n\,(P)H^{5/4}$, where P is the output power in kW, H is the effective head in metres and n is the rotational speed in revolutions per minute (rpm). The formula for the specific speed can be understood by considering a turbine scaled down to produce 1 kW of power at a head of 1 m for then its runner will rotate at the specific speed equal to its speed in rpm, i.e. $N_S = n$.

Figure 2.2 shows example efficiency curves for different turbine types. The speed of the runner blade relative to the water striking it is critical for the efficiency of the turbine. Propeller type turbines run best when their blade tips move faster than the water. In the case of the

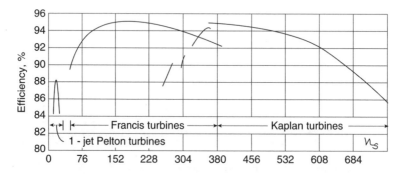

Figure 2.2   Efficiency curves for different turbine types

Kaplan, the blades should move at twice the water velocity, whereas the skirted Francis is most efficient when the two speeds are roughly equal. The Pelton theoretically performs best when its blades are moving at half the water velocity.

Reaction turbines use guide vanes at the rotor inlet to adjust the direction of the inlet flow and the flow rate in order to vary the shaft power in response to the electrical load. The turbine runner has a series of vanes whose complicated geometry is designed to extract maximum power under design conditions. These vanes cannot be moved instantaneously, which has implications for the power quality achievable from such turbines. For standalone applications where fast control is essential, electronic control of the loads is used to maintain turbine speed.

From Figure 2.2 it is clear that efficiency depends critically on specific speed, and this in turn depends on the effective head. Thus if the effective head varies significantly, the speed of the rotor needs to be adjusted to maintain high efficiency. Most turbines, however, operate at a fixed speed and efficiencies may thus vary to some extent.

Small hydro systems can be grid-connected by driving an appropriately sized induction generator (see Chapter 4) provided that there is sufficient flow control. Flows should be controlled by the guide vanes to limit the output of the generator to the rated value.

2.4 Wind Power

2.4.1 The Resource

Winds result from the large scale movements of air masses in the atmosphere. These movements of air are created on a global scale primarily by differential solar heating of the earth's atmosphere. Therefore, wind energy, like hydro, is also an indirect form of solar energy. Air in the equatorial regions is heated more strongly than at other latitudes, causing it to become lighter and less dense. This warm air rises to high altitudes and then flows northward and southward towards the poles where the air near the surface is cooler. This movement ceases at about 30 °N and 30 °S, where the air begins to cool and sink and a return flow of this cooler air takes place in the lowest layers of the atmosphere.

The areas of the globe where air is descending are zones of high pressure. Conversely where air is ascending, low pressure zones are formed. This horizontal pressure gradient drives the flow of air from high to low pressure, which determines the speed and initial direction of wind motion. The greater the pressure gradient, the greater is the force on the air and the higher is the wind speed. Since the direction of the force is from higher to lower pressure, the initial tendency of the wind is to flow perpendicular to the isobars (lines of equal pressure). However, as soon as wind motion is established, a deflective force is produced due to the rotation of the earth, which alters the direction of motion. This force is known as the Coriolis force. It is important in many of the world's windy areas, but plays little role near to the equator.

In addition to the main global wind systems there is also a variety of local effects. Differential heating of the sea and land also causes changes to the general flow. The nature of the terrain, ranging from mountains and valleys to more local obstacles such as buildings and trees, also has an important effect.

The boundary layer refers to the lower region of the atmosphere where the wind speed is retarded by frictional forces on the earth's surface. As a result wind speed increases with height; this is true up to the height of the boundary layer, which is at approximately 1000 metres, but depends on atmospheric conditions. The change of wind speed with height is known as the wind shear.

It is clear from this that the available resource depends on the hub height of the turbine. This has increased over recent years, reflecting the scaling-up of wind turbine technology, with the hub heights of the multimegawatt machines now being over 100 m.

The European accessible onshore wind resource has been estimated at 4800 TWh/year taking into account typical wind turbine conversion efficiencies, with the European offshore resource in the region of 3000 TWh/year although this is highly dependent on the assumed allowable distance from shore. A recent report suggests that by 2030 the EU could be generating 965 TWh from onshore and offshore wind, amounting to 22.6% of electricity requirements [4]. The world onshore resource is approximately 53 000 TWh/year, taking into account siting constraints. To see these figures in context note that the UK annual electricity demand is in the region of 350 TWh and the USA demand is 3500 TWh. No figure is currently available for the world offshore resource, and this itself will be highly dependent on the allowable distance from shore.

Of the new renewables wind power is the most developed. On very windy sites wind farms can produce energy at costs comparable to those of the most economic traditional generators. Due to advances in technology, the economies of scale, mass production and accumulated experience, over the next decade wind power is the renewable energy form likely to make the greatest contribution to electricity production. As a consequence, more work has been carried out on the integration of this resource than any of the other renewables and, naturally, this is reflected in the amount of attention given to wind power integration in this book.

2.4.2 Wind Variability

The wind speed at a given location is continuously varying. There are changes in the annual mean wind speed from year to year (*annual*) changes with season (*seasonal*), with passing weather systems (*synoptic*), on a daily basis (*diurnal*) and from second to second (*turbulence*). All these changes, on their different timescales, can cause problems in predicting the overall energy capture from a site (annual and seasonal), and in ensuring that the variability of energy production does not adversely affect the local electricity network to which the wind turbine is connected.

In Figure 2.3 each graph shows the wind speed over the time periods indicated. Wind speed measured continuously over 100 days is shown on the first graph followed by graphs, which in sequence zoom in on smaller and smaller windows of the series. It is easy to see the much larger relative variability in the longer time series (synoptic) as compared with the time series covering hours or less (diurnal, turbulence). This information is summarized in the spectral density presentation in Figure 2.4. In a spectral density function the height indicates the contribution to variation (strictly the variance) for the frequency indicated. A logarithmic scale as used here is the norm, and allows a very wide range of frequencies/timescales to be represented easily. The y axis is scaled by n to preserve the connection between areas under any part of the curve and the variance. The area under the entire curve is the total variance. It can be seen that the largest contribution to variation is the synoptic variation, confirming the interpretation of Figure 2.3. Fortunately these variations, characterized by durations of

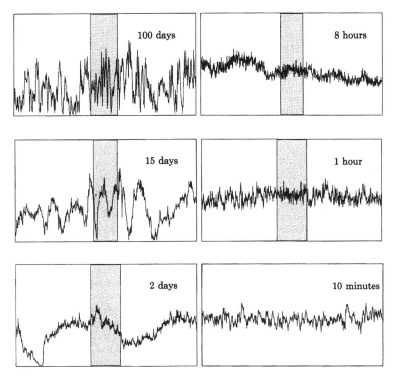

Figure 2.3   Wind speed measured 30 m above flat terrain: vertical axis is wind speed, 0–20 m/s. (Reproduced with permission of Risø National Laboratory for Sustainable Energy)

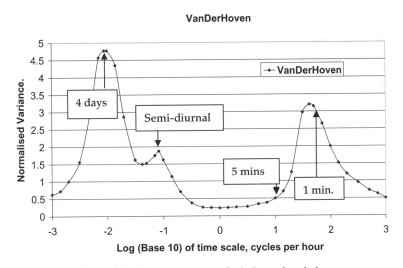

Figure 2.4   Power spectrum of wind speed variation

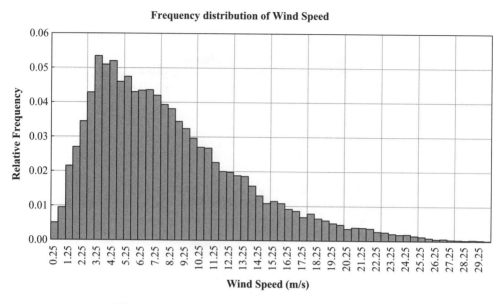

Figure 2.5   Example wind speed frequency distribution

typically 3 to 5 days, are slow in the context of the operation of large power systems. Apparently more difficult to deal with is the impact of short term variations due to wind turbulence, which are clear on the right hand side of Figure 2.4. However, as will be shown later, the aggregation effects will reduce this problem considerably. Fortuitously it is the timescales at which there is least variation, the so called spectral gap between 10 minutes and an hour or two, that pose the greatest challenge to power system operation.

The essential characteristics of the long term variations of wind speed can also be usefully described by a frequency or probability distribution. Figure 2.5 shows the frequency distribution for a year of 10 minute means recorded at Rutherford Appleton Laboratory, Oxfordshire, UK. Its shape is typical of wind speeds across most of the world's windier regions, with the modal value (the peak) located below the mean wind speed and a long tail reflecting the fact that most sites experience occasional very high winds associated with passing storms. A convenient mathematical distribution function that has been found to fit well with data, is the *Weibull* probability density function. This is expressed in terms of two parameters, k, a shape factor, and C, a scale factor that is closely related to the long term mean. These parameters are determined on the basis of a best fit to the wind speed data. A number of mathematical approaches of differing complexity are available to perform this fitting [5, 6].

2.4.3 Wind Turbines

The power in the wind than can be extracted by a wind turbine is proportional to the cube of the wind speed and is given in watts by:

$$P = \frac{1}{2}\rho A U^3 C_\mathrm{p}$$

Figure 2.6 The Vestas V90, 3 MW wind turbine. (Reproduced with permission of Vestas Wind Systems A/S)

where ρ is the air density, A is the rotor swept area, U is the wind speed and C_p is the power coefficient that represents the aerodynamic efficiency of the rotor. The variability in power output from one wind turbine would therefore be expected to substantially exaggerate the variability shown in the time histories of Figure 2.3.

Wind turbines are designed to generate their rated or nameplate output at a rated wind speed U_r. For wind speeds below a cut-in wind speed U_{co} the wind turbine is not operational as the developed aerodynamic torque is not sufficient to overcome the frictional losses of the drivetrain and generate a useful power. For wind speeds above rated the power is controlled aerodynamically to maintain the output at the rated value until some limiting wind speed value is reached, known as the cut-out wind speed U_{co} at which point the turbine is shut down. The relationship between power and wind speed is known as a power curve. A power curve for a 3 MW wind turbine illustrated in Figure 2.6 is shown in Figure 2.7. For this machine $U_{ci} = 3.5\,\mathrm{m/s}$, $U_r = 15\,\mathrm{m/s}$ and $U_{co} = 25\,\mathrm{m/s}$, values which are typical of large modern turbines.

This power characteristic combined with temporal variations in wind speed produces time varying electricity generation once the long term wind speed variations have been expressed in terms of a frequency or probability distribution of the sort shown in Figure 2.5. This can be combined with the power curve to indicate the probabilities of different power outputs,

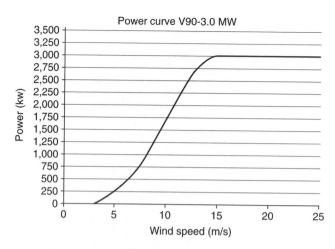

Figure 2.7   Power curve for the Vestas V90, 3.0 MW turbine. (Reproduced with permission of Vestas Wind Systems A/S)

and the overall average output of the turbine or wind farm. Details of this procedure can be found in Reference [7]. Average outputs lie typically in the range of 0.25–0.45 of the rated output depending on the mean wind speed at the site in question. These so called *load factors* or *capacity factors* are much lower than would be expected for conventional generators and are discussed in some detail in Chapter 3.

The aerodynamic manner in which the wind turbine rotor extracts energy from the wind is described in a number of textbooks [8–10]. There is a well defined upper limit to the aerodynamic efficiency C_p of a rotor; this is known as the Betz limit and is approximately 0.59. It reflects the fact that the air is not forced to flow through the rotor (e.g. as in a ducted turbine) but can flow around it instead. Conventionally C_p, is plotted as a function of the tip-speed ratio λ (defined by $\lambda = \Omega R/U$ where Ω is the angular velocity of the rotor, R the radius and U the incident wind speed). The ratio λ is defined in this way to provide a generalized representation of the wind turbine rotor performance that is applicable to all combinations of incident wind speed and rotor rotational speed.

A typical C_p–λ characteristic is shown in Figure 2.8. It is apparent that to operate at peak efficiency (for large modern wind turbines C_p is normally in the range 0.4–0.5), the tip speed ratio must be held constant for maximum output and this requires the rotor speed to be controlled in proportion to the wind speed. This is one reason why most larger modern wind turbines are designed to operate at variable speed (see Chapter 4 for a more complete discussion). This is attractive from an integration perspective as the rotor has inertia available to absorb or release energy when accelerating or decelerating respectively, thus smoothing short term variations in wind speed. Consequently its electrical power output varies less and can be more easily accommodated by the electrical system.

The C_p–λ of Figure 2.8 assumes a fixed blade configuration. If the blade orientation is changed, then the efficiency will change too. In fact if the blades are *pitched* or *feathered*, i.e. rotated about their axis so as to reduce the angle between the blade chord and the resultant incident wind, then the lift forces on the blade that produce the torque will reduce. This will

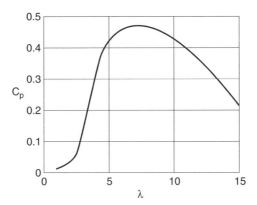

Figure 2.8 C_p–λ curve for a three-bladed rotor

result in a C_p–λ curve that sits below the one shown in Figure 2.8. The more the blades are pitched away from the optimal position, the lower the output at any given combination of wind and rotor speed. Wind turbines that can control their output by pitching the blades in this manner are called *pitch-controlled*. An alternative is to allow the process of stall to limit the wind turbine output. Such *stall regulated* machines have a fixed blade pitch and run at a nominally fixed rotor speed so that wind speeds above rated result in stalling of the blades, thus limiting power output around the rated value. Stall regulation was widely used for the smaller turbines (50–500 kW), but this passive approach to power control lacks the potential for operators to adjust the turbine output in response to external conditions. In part as a consequence of this, practically all MW sized wind turbines are now pitch-controlled in some way. An approach known as *active stall control* combines pitch and stall regulation. Dynamic power control is by stall but the blades are slowly adjusted to make sure stall occurs at the correct power level.

The mechanical shaft power created by the wind turbine rotor is converted to electricity by an electrical generator with a conversion efficiency that may reach 98% for large generators. Variable speed operation requires a frequency conversion through a power electronic converter, a process that reduces somewhat the overall efficiency. The generator and frequency conversion aspects are discussed in Chapter 4.

2.4.4 Power Variability

The short, medium and long term variations in wind speed shown in Figure 2.3 affect power system operation in different ways. This is sufficiently important, given the significant and growing proportion of wind capacity in a number of power systems, to merit more detailed discussion.

Variability from Second to Second

At low/medium wind speeds the electrical output from a single wind turbine could vary substantially. In Chapter 5 it will be shown that this may have a detrimental effect on the power system. However, when wind turbines are clustered in wind farms, there is physical spacing between them and the turbulence seen by each wind turbine is different and to a great extent uncorrelated. The electrical output from wind farms therefore exhibits substantially lower relative variability than that from a single wind turbine. At the planning stage, appropriate analytical studies are carried out to ensure that the variability expected from a wind farm at a particular site will not adversely affect the power system.

Variability from Minute to Minute

Figures 2.3 and 2.4 indicate that the character of wind is such that if the second to second turbulence is removed, the average wind speed from 10 minute period to 10 minute period remains effectively constant. In Chapter 3 will be shown that this 'persistence' nature of such averaged wind speeds is particularly important in integrating wind generated electricity in power networks. In practice the output of turbines can be regarded as uncorrelated on the timescale of minutes and as with the faster variations the affect of aggregation is to smooth out variations at these higher frequencies.

Variability from Hour to Hour and from Day to Day

Figure 2.9 shows the actual records of wind speed at 13 geographically dispersed wind sites in the UK. As expected from Figure 2.4, there is substantial variability at each location over

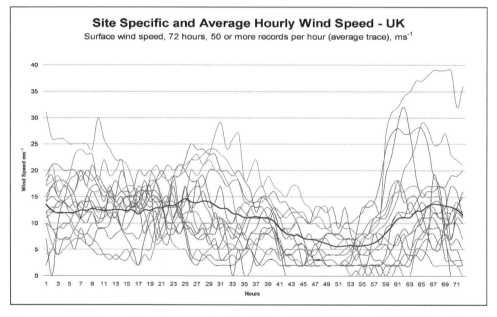

Figure 2.9 UK site specific and average hourly wind speed over 72 hours, with 50 or more records per hour. (Reproduced from Sinden, G.E., 2007, DPhil Thesis with permission of Environmental Change Institute, Oxford University Centre for the Environment)

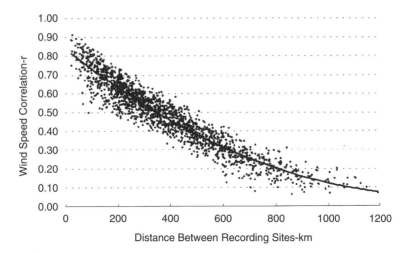

Figure 2.10 UK onshore wind power correlation by distance between sites based on UK long term averages. (Reproduced from Sinden, G.E., 2007, DPhil Thesis with permission of Environmental Change Institute, Oxford University Centre for the Environment)

the 72 hour recorded period. Variability now becomes significant and effective integration of wind power in an electrical power network would have been problematic if the *total* resource were to fluctuate in sympathy with the fluctuations in one site. Fortunately this is not the case due to the geographic diversity of the sites. The benefits of geographic diversity are clearly illustrated by the bold line in Figure 2.9, which represents the average wind speed from all sites. This average is substantially smoother (i.e. exhibits much reduced variability) than the wind at the individual wind farm sites.

With large scale exploitation of the wind resource, wind farms are installed inevitably across a variety of geographically dispersed sites. As suggested above, this has a major beneficial effect in terms of overall variability. Just as the output from a wind farm has less short term variability than a single wind turbine due to their dispersion across the site, so the aggregate output from several geographically dispersed wind farms has less longer term variability than the output from a single wind farm. This reflects the fact that distant localities experience variations in wind due to shifting weather patterns that are time shifted in relation to one another, and also to an extent distinct. Figure 2.10 presents the correlation between pairs of onshore wind sites in the UK as a function of the distance between the sites, and demonstrates that sites very far apart exhibit low cross-correlation. The data from Reference [11] were recorded over a period of 15–20 years.

Seasonal Variability

Seasonal and monthly average wind speeds vary significantly over most of the world. Figure 2.11 shows the seasonal changes of monthly averaged wind speeds for Billings, Montana in the USA. The trend of higher wind speeds during winter compared to summer is typical of the Northern hemisphere. The figure also indicates that there is variability from year to year.

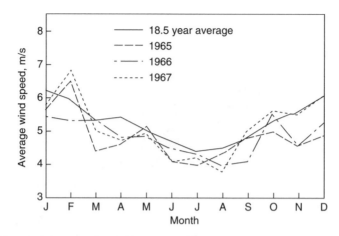

Figure 2.11 Seasonal changes of monthly average wind speeds. (Reproduced from Reference [10] with permission of John Wiley & Sons, Inc.)

2.5 PV and Solar Thermal Electricity

2.5.1 The Resource

The average intensity of light outside the atmosphere (known as the solar constant) is near to 1353 W/m². Attenuation by the atmosphere results in peak intensity at sea level of around 1 kW/m², giving a 24 hour annual average of 0.2 kW/m² averaged over the planet's surface. As this overall energy density is relatively small, large areas will be needed for a significant energy production. For example, in order to produce a gigawatt of power, an area of nearly 5 km² would be needed (assuming a conversion efficiency of 20%). However, in countries such as the UK (with low irradiance[7] and high population density) the existing and appropriate roof space is sufficient in principle to generate enough electricity to cover a significant proportion of electricity consumption.

Common sense tells us that irradiation varies regionally, with the changing seasons, and hourly with the daily variation of the sun's elevation. Many locations do not experience unbroken sunshine. Cloud cover can significantly reduce the net radiation and cause relatively fast variations in intensity, in some cases significant variations from minute to minute or even over seconds.

Seasonal variation on the earth's surface is illustrated here by comparing two different sites. The magnitude will obviously vary from site to site but the basic principles are the same. Sites of similar latitudes should, in principal, have a similar solar resource.

The earth's tilt angle leads to a variation of the seasonal irradiance. Figure 2.12 illustrates the differences in average monthly irradiation for a site close to the equator, Kisangani (Congo, latitude –0.31 °) and a site at higher latitude, Sutton Bonnington (UK, latitude 52.5 °). As expected, Kisangani receives twice as much irradiation in the course of the year. While

---

[7] Sometimes the terms solar radiance, radiation intensity and insolation (in older text books) can be found although irradiance is the most widely accepted. The SI units for irradiance are W/m².

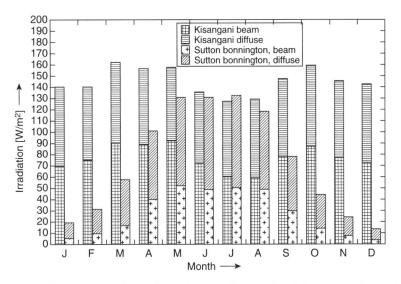

Figure 2.12   Comparison of Irradiation at Kisangani (Congo) and Sutton Bonnington (UK)

Sutton Bonnington experiences a strong seasonal variation of ±73%, the irradiation in Kisangani varies only by ±13% around the yearly average.

The annual average irradiation peaks in the north-east of Africa at around 300 W/m². Other favoured regions can be found all over Africa, the north of South America and the south of North America, and South-West Asia. Australia experiences solar irradiation levels above 250 W/m² across nearly half the continent. The irradiation reduces, generally speaking, the further the distance away from the equator. It can be well below 100 W/m² on a 24 hour average. Obviously, these values are averaged over the year and will vary significantly with the seasons. Most northern European, North American and north Asian countries are within this latter region and present and future systems are and likely to be rooftop installations in the kilowatt scale connected to the local 230 or 115 V network.

2.5.2 The Technology

There are two main technologies for the conversion of sunlight into electricity. Photovoltaic (PV) cells depend on the use of semiconductor devices for the direct conversion of the solar radiation into electrical energy. Efficiencies of the typical commercial crystalline PV cells are in the range 12–18% although experimental cells have been constructed that are capable of over 30%. In contrast, solar thermal systems depend on intermediate conversion of solar energy into thermal energy in the form of steam, which in turn is used to drive a turbogenerator. To obtain high temperatures, thermal systems invariably use concentrators either in the form of parabolic troughs or thermal towers. At present, generation of electricity by either technology is substantially more expensive than traditional means. Due to the considerable potential of cost reductions in PV systems it is believed that in the future, perhaps in a decade or so from the time of writing, PV systems will be providing a sizeable proportion of the renewable energy contribution.

Some countries, notably Japan and Germany, have created substantial home based markets for PV cells, and the last couple of years have seen a rapid growth in multimegawatt installations in Europe. Despite this encouraging trend, the relatively high present cost of PV systems are likely to limit them to modest contributions to overall electricity supply in the immediate future. In sunnier places such as California, Australia, North Africa and the Mediterranean, where the peak electricity demand occurs in summer due to tourism and air-conditioning loads, large scale multimegawatt PV plants or solar thermal plants are more attractive. Nevertheless, significant technology and cost breakthroughs in these two solar technologies will be needed if they are to make sizeable contributions to electricity supply.

The characteristics of PV and solar thermal plants are rather different and are next outlined.

2.5.3 Photovoltaic Systems

At the heart of a PV system is the PV module. Detailed descriptions of the different PV technologies and the basics of solar cell operation can be found in a wide range of textbooks, for example References [13] and [14]. PV modules produce output determined mainly by the level of incident radiation. They are characterized for given external conditions, by an *I–V* curve of the type shown in Figure 2.13. The power, *IV*, depends on the operating point and is maximized for operation near to the knee of the *I–V* characteristic, known as the maximum power point (MPP). Power electronics is used to convert the DC (direct current) output of the PV modules to AC (alternating current) for injection into the network (more about this in Chapter 3). The quality of a cell can be judged by the squareness of the *I–V* characteristic. This is quantified in terms of the ratio of the voltage at open circuit (i.e. where the *I–V* curve meets the voltage axis) times the closed circuit current (i.e. where the *I–V* curve meets the current axis), divided by the power at the MPP. This ratio is known as the fill factor.

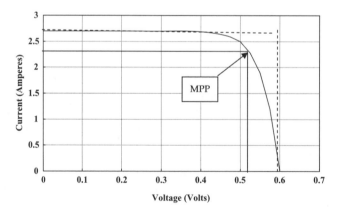

Figure 2.13   An example *I–V* curve

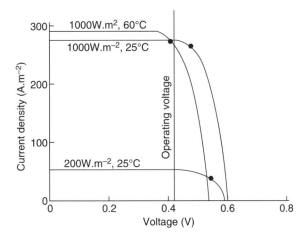

Figure 2.14 Impact of radiation and temperature on the *I–V* characteristic

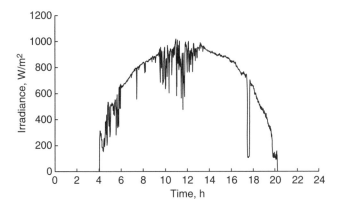

Figure 2.15 Time variation of radiation through a summer's day in Loughborough, UK

As shown in Figure 2.14, radiation and temperature affect the shape of the *I–V* curve and the voltage at which the MPP occurs. The power electronic converter is also used to control the operating voltage as near as possible to the MPP and track this as it changes with radiation level and to a lesser extent with module temperature. Commonly both of these power electronic functions are combined in the inverters used for grid connection. PV modules have negligible energy storage capability, and as far as the power system is concerned their output varies instantaneously with changes in radiation. Hence a passing cloud, as occurs in Figure 2.15 just before 18.00, will result in the sudden collapse in output from a PV system. As with wind turbines, a geographical spread of PV systems will substantially mitigate such short term effects.

It has been speculated that as power systems become increasingly interconnected, as is currently happening in Europe, it may eventually be possible to smooth out the diurnal

Figure 2.16   50 MW parabolic trough plant with thermal storage at Andasol, Spain. (Reproduced from Concentrating Solar Power – from research to implementation. © European Communities, 2007)

variations in PV output.[8] In such a vision significant intercontinental power flows might occur; for example, power generated from massive PV arrays in the Sahara during hours of daylight could be transmitted to areas of the planet then in darkness. Of course this would involve power systems spanning substantial proportions of the planet, something that may never turn out to be cost effective.

2.5.4 Solar Thermal Electric Systems

Solar thermal electricity generation systems most commonly use solar concentrators to produce high temperatures that can drive heat engines with acceptable conversion efficiency. Two main types of generator suitable for large scale generation have so far been demonstrated, both requiring direct or beam radiation. Climates with cloudy or overcast conditions, where most of the radiation is diffuse, are thus unsuited to this technology. The two technologies are:

- Large collections (or solar farms) of parabolic trough reflectors, Figure 2.16, focus solar radiation on to a line receiver containing a piped heat transfer medium. The medium can be a thermal oil, capable of withstanding high temperatures, or alternatively pressurized water. This is collected and passed through a heat exchanger/boiler where steam is raised for supply to the turbines. Operational temperatures vary between 350 and 400 °C and system sizes up to 80 MW have been built. To increase steam turbine operating

---

[8]The benefits of interconnection are central to the operation of power systems and will be returned to again and again throughout this book.

Figure 2.17 Aerial view of 11 MW PS10 solar tower plant at Sanlúcar la Mayor, Spain. (Reproduced from Concentrating Solar Power – from research to implementation. © European Communities, 2007)

temperature, and thus efficiency, steam from the solar system can be further heated using conventional fuels.

- In solar power towers a central receiver is mounted on top of a tower which is surrounded by a field of concentrating mirrors (heliostats) which track the sun. The heliostats reflect and concentrate the light on to a receiver where the energy is absorbed by the heat transfer medium, which could be water, a molten salt or any other suitable high temperature heat transfer liquid. Temperatures as high as 1000 °C can be achieved. System sizes up to 200 MW have been considered, but to date the largest systems constructed are considerably less. Figure 2.17 shows one of the larger systems currently operating, the 11 MW PS10 solar tower plant at Sanlúcar la Mayor, Spain. Solar power towers are expected to be more economic than solar farms for larger sized plant, say over 100 MW.

Since both these technologies depend on direct sunlight for efficient operation they function for only part of the time, and then not always at maximum output. However, because they both involve a thermal intermediary stage, they can be combined with fossil fuel combustion (a hybrid) and depending on design details it may be possible to incorporate thermal storage.

Table 2.1 Performance of parabolic trough and power tower systems. (Source: http://www.eere.energy.gov/consumerinfo/pdfs/solar_overview.pdf)

	Parabolic trough	Power tower
Size	30–320 MW	10–200 MW
Operating temperature (°C/°F)	390/734	565/1049
Annual capacity factor	23–50%[a]	20–77%[a]
Peak efficiency	20%(demonstrated)	23%(predicted)
Net annual efficiency	11(demonstrated)–16%	7(demonstrated)–20%
Status	Prototype	Demonstration
Storage available	Limited	Yes
Hybrid designs	Yes	Yes

[a]The top end capacity factors are for systems including substantial thermal storage.

The primary advantage of these adaptations is to the capability to supply power when the sun does not shine, in other words to be dispatchable.

A smaller scale approach, suitable for units in the range 10–50 kW, is possible using parabolic concentrators. These also work at high temperatures of 600–1000 °C and might be suitable for distributed electricity generation.

Thermal-to-electric efficiency is in the range of 20–40% depending on the design, with the resulting overall solar-to-electricity conversion efficiency in the range 13–25%.

Table 2.1 (adapted from material available from the website given in Reference [14] when it was accessed in January 2006) compares the performance of the parabolic trough and power tower systems.

2.6 Tidal Power

2.6.1 The Resource

The moon and sun's gravitational fields cause the natural rise and fall of coastal tidal waters. Since the moon is closer to the earth, albeit much less massive, it has a dominant effect upon tides. As the moon is 2.2 times more influential than the sun, it could be considered that tidal energy is mostly a form of lunar energy!

The earth rotates on its axis once every 24 hours. In the earth's frame of reference the sun orbits the earth once every 24 hours. The moon orbits the earth once every 29 days approximately. In the earth's frame of reference, the moon appears to orbit the earth once every 24 hours and 50 minutes. This difference in periods between the apparent orbits of the sun and moon leads to phase changes with larger *spring tides* during in-phase behaviour and smaller *neap tides* when the sun and moon are out of phase. Spring tides can be twice as large as neap tides.

A 1991 study commissioned by the EU estimated that the technically feasible energy resources from tidal barrages across the EU could be as much as 105 TWh/year (from 64 GW of installed capacity). This resource is unevenly distributed across Europe with the UK

(47.7%) and France (42.1%) sharing the bulk of the resource and Eire (7.6%) accounting for most of the rest.

It has been estimated that exploitation of all practicable estuaries in the UK could lead to electricity generation of up to around 20% of demand in England and Wales.

2.6.2 Tidal Enhancement

Over much of the surface of the oceans, the *tidal range* (the vertical rise and fall) is rather small, less than one metre, but in certain places, there is an enhancement of the range. Enhancement may be due to the following.

Funnelling

The tide is gradually constrained from the sides and so increases in height – or the reverse, later in the cycle.

Resonance

The estuary has a resonant period equivalent to the tidal period. The length and depth of the estuary are very important for resonance.

Coriolis Effect

The spinning of the earth leads to the Coriolis effect mentioned earlier with respect to wind. The tide is influenced by this and in some locations tends to increase in height at high tide and be drawn away from the coast at low tide, with the net effect of enhancing the tidal range.

These enhancements are highly predictable, allowing the output from a tidal scheme to be determized many years ahead.

2.6.3 Tidal Barrages

Constructing a barrage across an estuary and allowing tidal waters alternately to fill the estuary through sluice gates and then to empty it through turbines can generate energy. A barrage constructed across an estuary is equipped with a series of gated sluices and a bank of low head axial turbines. Where it is necessary to maintain navigation to the upper part of the estuary, a ship-lock may be required.

Tidal barrages are a currently available technology, but very few exist worldwide. The best known example is the 240 MW scheme at La Rance in France, and smaller installations have been made in Nova Scotia, Russia and China. The UK has a number of attractive sites due to its high tidal ranges, the largest potentials being on the Severn estuary (8600 MW capacity) and the Mersey (700 MW capacity). However, the scale of these installations and the calculated long payback periods make the required investments unlikely in the context of the privatized electricity supply industry. Furthermore, environmental considerations may present a barrier to large scale developments.

2.6.4 Operational Strategies

Ebb generation is the simplest mode of operation for a tidal barrage scheme. The operating cycle consists of four steps:

- Sluicing on the flood tide, to fill the basin.
- Holding the impounded water until the receding tide creates a suitable head.
- Releasing the water from the basin to the sea via turbines, on the ebb tide, until the tide turns and rises to reduce the head to the minimum operating point.
- Holding until the tide rises sufficiently to repeat the first step.

Ebb generation with flood pumping is a modification of this mode which allows increased energy output. By using the turbines in reverse as pumps, the basin level and hence the generating head can be raised. The energy required for pumping must be imported but since the pumping is carried out against a small head at high tide and the same water is released later through the turbine at a greater head, this can produce a net energy gain with some limited ability to re-time output. The energy gain through pumping could be small but useful and typically in the range 3–13%.

Flood generation is the reverse of ebb generation and is rarely suggested alone, possibly because it offers little storage opportunity as the basin is often being filled by a river, which reduces the total energy capacity.

Two-way generation (ebb and flood) is possible with reversible turbines and is used at La Rance (together with flood pumping). The additional energy recovered may not justify the extra cost and complexity of the turbines.

Most schemes propose ebb-only generation. The full resource could only be captured if the barrage of the scheme has a very large two-way generating flow capacity in order to achieve rapid filling and emptying of the basin. In practice, it is neither economic nor is there enough physical space to install enough turbines to come anywhere near full resource capture. Typically only about one-third of the gross power can be harnessed, so after turbine and generator losses the mean electrical output will at best be about a quarter of the gross power.

The other main reason for preferring ebb generation is the nature of the amenity that results. In such a scheme, the water level in the basin never falls below the mean sea level, so the basin is available for recreational activity much like an inland reservoir. On the other hand, it is bad news for the bird population for whom exposed mudflats at low tide provide a major food resource. High initial capital cost and polarized opinions on the resulting environmental changes are the main reasons why tidal schemes seldom get beyond the feasibility study stage. Tidal energy barrages are expected to have very long lifetimes. Their design life could be about 120 years, but with normal maintenance and replacement of turbine generators at 40 year intervals, their lifetime could effectively be unlimited.

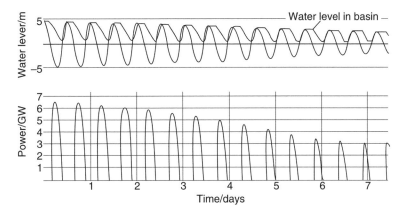

Figure 2.18   Water level and electrical power output of the proposed Severn Barrage over a spring–neap tide cycle. (Source: the Watt Committee on Energy now disbanded. Reproduced with permission of Oxford University Press)

Power Variability

One of the main disadvantages of tidal schemes is the pulsed nature of their electrical output. Figure 2.18 shows the expected electrical output from the 7 GW scheme proposed for the Severn estuary assuming ebb generation only.

Power is generated for five to six hours during spring tides and for three hours during neap tides. This pattern may, but is unlikely to, match the pattern of demand in an integrated power system. On the positive side the power availability from such a scheme is highly predictable. In a large integrated system such as the UK's this predictability will ease the task of scheduling alternative generation to take up the demand when the output from the scheme falls to zero.

2.6.5 Tidal Current Schemes

The direct production of electricity from tidal streams is a relatively new concept. It relies on a different approach to conventional tidal barrage schemes and is attractive in that it does not require such massive infrastructural investment.

Tidal stream technology extracts energy directly from the currents that flow in certain locations, driven by the rise and fall of tides in the vicinity. These currents usually have a low velocity (1 m/s), though this can be enhanced by the local topography. In particular, the velocity can be magnified greatly in straits between islands or between islands and the mainland. Tides can be predicted with very high accuracy; hence after measurements at a site, the energy available for conversion can be forecast with confidence.

Figure 2.19 shows a tidal cycle together with the associated stream current velocity. Tidal height varies approximately sinusoidally and, as already mentioned, is more or less completely predictable. For any site the tidal current velocity has four peaks and four troughs per day, as in the figure.

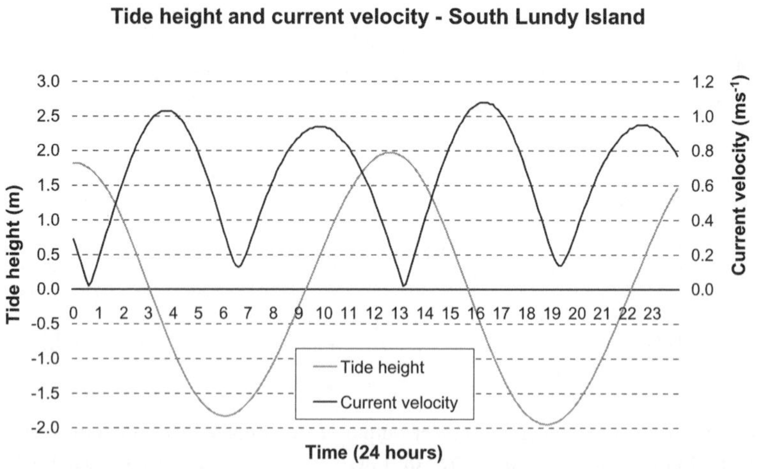

Tide height and current velocity - South Lundy Island

Figure 2.19 Tide height and current velocity. (Reproduced from Sinden, G.E., 2007, DPhil Thesis with permission of Environmental Change Institute, Oxford University Centre for the Environment)

The technology to extract energy from the tidal current is conceptually simple: a turbine is placed in a suitable tidal flow, which turns the generator through a gearbox as in Figure 2.20. It is similar to a submerged wind turbine, except that the greater specific gravity of seawater results in much higher energy densities in tidal streams than is found in winds of the same velocity. However, the water velocities available in tidal streams (typically rated velocities of 2–3 m/s on good sites) are much lower than the air velocities used by wind turbines. Although power output is proportional to the cube of the velocity, tidal stream rotors generally produce significantly greater output than wind turbines of the same size because of the massively increased water density. Compared to wind, tidal flow velocities are expected to have little turbulence and thus vary in a smooth manner, reducing fatigue loads on the rotor and generating electricity with little short term variation.

At different sites, the peaks and troughs occur at different times. Power being proportional to the cube of current velocity, this implies that maximum and minimum electricity generation is possible at different times. By combining generation from sites that are out of phase it is theoretically possible to smooth out the diurnal variability. The tidal stream resource is highly site-specific, and there is a limited number of tidal stream sites around the world worth exploring. The accessible tidal stream resource for the most suitable sites in the UK (including the Channel Islands) is estimated to be approximately 36 TW h/year. The power variability from such schemes will be similar to that from barrages.

A second type of variability also affects tidal velocities. This is the spring–neap cycle, which occurs over 14 days and changes tide ranges (the difference in height between high and low tides) from a minimum to a maximum, then back to a minimum. Unlike diurnal variations, the spring–neap cycle affects all sites in the same way at the same time, so there is no scope for smoothing by combining generation from many sites.

It is concluded that patterns of tidal stream power output differ significantly from electricity demand cycles; maximum potential output sometimes coincides with peak demand but at other times minimum output does. However at the likely levels of penetration

Figure 2.20 Artist's impression of axial flow tidal current turbines. (Reproduced with permission of Marine Current Turbines Limited)

from this resource, the magnitude of this variability is small compared to demand variations.

2.7 Wave Power

2.7.1 The Resource

The passage of wind over the surface of the sea results in the gradual transfer of energy into the water to produce waves, so wave energy is also an indirect form of solar energy. Wind power typically has densities in the range 1.2–1.8 kW/m². Waves with a typical power density of 50 kW per metre of wave front or crest length are in effect a highly concentrated form of solar energy. The distance over which this process of wave generation occurs is called the 'fetch' and longer fetches produce larger, more powerful waves as do stronger winds and extended periods of wind. Sea waves are characterized by their wave height (H), period (T) and crest length. The power per metre of crest length is proportional to the period and to the square of the wave height.

Wave records indicate that the heights and periods of ocean waves vary continuously over time, which results in wave power varying almost continuously. However, the conversion of the energy of a wave face into electricity output tends to partly smooth this variability. Because of these irregularities the calculation of the power density requires an averaging process. The *significant wave height* H_S is approximately equal to the average of the highest

Table 2.2   Technically achievable wave power resource

	UK potential (GW)	World potential (GW)
Shoreline	0.03	1–50
Near shore	0.3–0.7	10–500
Offshore	7–10	200–5000

one-third of the waves and the *zero up-crossing period* T_e is defined as the average time counted over ten crossings or more between upward movements of the surface through the mean level. It can be shown that the average power in one metre of wave crest is proportional to T_e and to the square of H_S.

The world wave resource is not yet fully analysed but there is no doubt that it is sizeable. Table 2.2 gives an estimate of the scale of the technically achievable resource.

2.7.2   The Technology

The conversion of wave power into electricity requires a device that intercepts the waves and converts a proportion of their energy first into mechanical and then into electrical form. The conversion of wave energy into mechanical energy demands a central stable structure incorporating an active element which moves relative to it under the forces exerted by the waves, and can react against the central structure to produce forces and displacements that generate mechanical power.

Wave energy research and development has been taking place for over 25 years now, and significant progress has been made towards the development of viable technologies able to exploit the large energy potential of the world ocean wave climates. Ocean waves are often powerful, but with extremely low frequencies, of about 0.1 Hz (equivalent to 6 rpm or to periods of around 10 s), and the success in generating electricity demands that this frequency is raised to 500–1500 rpm.

The technology for wave energy conversion is still at an early stage compared to wind and photovoltaics and a variety of different approaches are currently under development. Because of the nature of the resource and for efficient conversion, the swept volume of the device must be of the order of several tens of cubic metres per metre of device width. The devices are therefore physically large and have to be designed to withstand without damage extreme waves that may occur very rarely. The transition of a concept from a model tested in a laboratory wave tank to a working prototype requires considerable expenditure and it is only very recently that serious funding from governments and industry has been forthcoming.

It is not the intention here to describe in detail the various concepts and associated hardware. The devices can be classified in a variety of ways depending on their intended location or their geometry and orientation. At least twelve different devices are being developed worldwide and it is still too early to guess which of them will be capable of providing energy competitively and reliably. As an example of the ingenuity of recent developments, Figure 2.21 shows an artist's impression of a Pelamis array. The structure of the Pelamis comprises four semi-submerged cylinders linked by hinged joints. The relative motion between the

Figure 2.21 Artist's impression of Pelamis array. (Reproduced with permission of Pelamis Wave Power)

cylinders is resisted by hydraulic rams that pump oil into high pressure tanks which then is used to drive an electrical generator.

While the UK is fortunate in having a good wave climate, the political climate has not always favoured the technology. Attitudes, however, are changing, prompted by the need to address global climate change, by the long term resource security of fossil fuels and by the increasingly competitive economics of wave energy. UK R&D teams conducted much of the early work, but many other countries are now active in wave energy development. Commercial involvement is now significant and schemes or concepts from the Netherlands, Norway, Australia, Sweden, Denmark, the USA as well as the UK are now being developed.

2.7.3 Variability

The power output from a single wave power device will follow an exaggerated variation of the wave height trains because of the square relationship between power and wave height. Not unlike offshore wind farms, a substantial wave power installation will consist of an array of such devices, the outputs from which when added together, because of spatial effects, will be relatively smoother than that from one device. If this were extended to hundreds of such devices in geographically dispersed arrays, the overall output will be smoother still. The short term variability of the output from wave power converters is unlikely to present integration problems, assuming that the outputs from arrays are connected to a grid voltage level capable of absorbing variable generated power without adverse network effects.

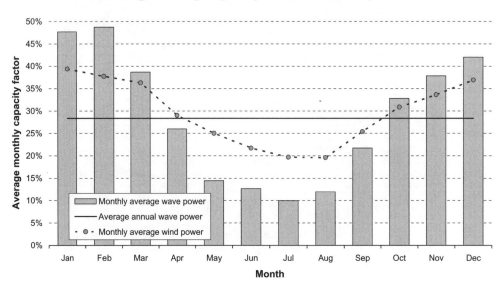

Figure 2.22 Average monthly wave power in the UK. Average monthly wind power capacity factor [15] is shown for reference purposes. (Reproduced from Sinden, G.E., 2007, DPhil Thesis with permission of Environmental Change Institute, Oxford University Centre for the Environment)

The wave resource, not unlike the wind resource on which it depends, also varies on a day-to-day and season-by-season basis; in general wave conditions are more energetic in the winter than in the summer. For example, about half of the annual wave power at all UK sites occurs during the winter months of December, January and February, as shown in Figure 2.22. This exceptional match between availability and demand for the UK would also hold for many countries with coastal areas facing the Atlantic or Pacific in the northern hemisphere. Figure 2.22 also depicts the seasonal variability of wind to illustrate, by comparison, the pronounced variability of wave power.

Wave power levels [16] are predictable to varying degrees over different timescales. At short forecast periods (a few hours), information on the wave conditions offshore, coupled with the power available from wave devices in the previous hour, can be used to estimate future output. For longer forecast periods, wave models provide estimates of wave power up to 5 days ahead.

2.8 Biomass

2.8.1 The Resource

Biomass differs considerably from the other renewable sources in that it takes the form of a fuel that can be stored and used for electricity generation when required, in the same way as fossil fuels. However, unlike fossil fuels, biomass is often limited by the energy density of the stored fuel. Therefore, it must be produced and consumed locally, as energy consumption associated with transportation over long distances might even exceed that of the fuel itself. This means that biomass power generating units are relatively small compared to conventional plant, (relying on local supply chains for feedstock) and possess the characteristics of small embedded generating units.

There are three basic thermochemical conversion technologies that use solid biomass as a primary fuel for the production of electricity, namely direct combustion, gasification and pyrolysis. In addition, the use of liquid biomass (such as sewage sludge) for the production of methane via anaerobic digestion is increasingly common.

Electricity production using solid biomass fuels is still a developing industry and as a consequence is not competitive on price with electricity from fossil fuels without some kind of government fiscal or policy support. However, it is competitive with nuclear power and possibly new-build clean coal power stations, but not with modern gas fired power stations within the current regulatory and economic climate. With the correct support, as currently in the UK, co-firing of coal with biomass is commercially attractive. In the longer term, grid-connected biomass generation (using the full range of possible technologies) may become competitive; the greatest potential is for small scale embedded generation using gasification, pyrolysis or high speed steam engine based plant. In the short term, small scale (100–500 kWe) dedicated plants for use on farms or by rural industry has the greatest potential. In the medium term when increased demand for electricity could be causing the grid to become overloaded and unreliable, then larger (1–20 MWe) embedded biomass generation plant providing end-of-grid support may become an attractive alternative to reinforcing the grid.

2.8.2 Resource Sustainability

Biomass has become less important as countries have industrialized and now accounts for less than 3% of energy in the developed world. In contrast developing countries remain highly reliant on wood and other natural biomass with over 30% of their energy needs being supplied from these sources. Of course with growing populations this supply is not sustainable. Ironically, the industrialized countries need to make increased use of biomass, and the less developed regions limit their use of this resource to a sustainable level.

One of the key features of biomass is that the energy expended in growing it, i.e. planting, watering, use of chemicals and pesticides to enhance yield, harvesting, drying etc., is not negligible. Specifically for ethanol production used as fuel for transport, the refineries themselves are fired by fossil fuels to ferment the crop and to purify ethanol from the product of fermentation. A US Department of Agriculture report found that the energy from corn bio-ethanol was a mere 8% in excess of the input production energy and a recent paper in *Science* [15] found that the energy ratio was net-positive when the energy savings from 'co-products' for cattle feed were included. Efforts are now being made to produce bioethanol from cellulosic crops and not from fermentation. This promises to produce twice the amount of ethanol per hectare of crop.

The scene is a lot brighter if biomass is used to generate electricity, especially in CHP plants in small decentralized power stations. The benefits are compounded if the crops are grown organically, if possible, and used locally. The choice of crop is also vital in the effectiveness of CO_2 mitigation. Table 2.3 compares the ratio of energy out to energy in for a

Table 2.3 Energy ratio of several crops

Crop	Ratio
Miscanthus	32.5
Willow	30.0
Hemp (straw)	8.5
Wheat (grain)	8.8
Oilseed rape	3.8

number of crops based on DTI report URN 01/797, [16]. Miscanthus and willow are the preferred crops in European and other similar latitudes. Currently, the main use of miscanthus in the UK is in co-firing with coal in existing power stations.

Large scale biofuel production is not only energy intensive but it could have an adverse environmental and social impact. Such development requires substantial water resources with the result that water tables in areas of intense cultivation have been lowered to unacceptable levels. Expansion of biofuel crops could also speed up tropical deforestation with the associated lowering of CO_2 absorption and threat to extinction of thousands of species of animals and plants. If such crops are encouraged through subsidies, food shortages may occur if land previously used to produce food is lost. It could be concluded that the biomass path, unless used wisely, may cause serious environmental impacts.

2.9 Summary of Power Generation Characteristics

This chapter has reviewed the main forms of electricity generation. For the renewable energy forms that are the main topic of this book, the resource characteristics have been presented and discussed in some detail since they impact significantly on the manner in which these sources can be integrated into the power system. This chapter concludes by summarizing the key characteristics of the different forms of generation in Table 2.4.

In this table typical unit sizes for the different forms of electricity generation are referred to. These unit sizes are in all cases the nameplate rating of the prime mover/generator package. Power stations may contain one or several such units. The table also describes the nature of the energy resource in terms of availability. Traditional generators supplied from coal, oil, gas or nuclear energy score highly here as the primary energy resource is continuously available. However, their capability to be dispatched, which requires that their output can be changed automatically or at will by control engineers to follow the demand variations, depends crucially on their thermal/mechanical nature and differs from plant to plant as described in Chapter 3.

One characteristic shared by all the renewable sources, excluding biomass and tidal, is their variability and relative unpredictability. This presents a challenge in integrating such sources in electrical power networks that have been designed to operate with traditional generators whose availability appears certain. In reality, however, no plant can be completely available; there is always some probability of breakdown, the need for maintenance, etc. This topic is complex and often misunderstood. It is dealt with in some detail in Chapter 3.

Table 2.4 Generator characteristics by energy source

Energy source	Typical unit size	Variable[a]	Predictable	Dispatchable
Coal	500 MW	No	Yes	Yes
Nuclear	500 MW	No	Yes	No
Gas CCGT	Up to 500 MW	No	Yes	Yes
Gas open cycle	100 MW	No	Yes	Yes
Hydro with reservoir	Up to 500 MW	No	Yes	Yes
Pumped storage hydro	Up to 500 MW	Yes	Yes	Yes
CHP	Up to 100 MW	Usually	Usually	No, because it is heat led
Energy crops and municipal solid waste	Up to 40 MW (at present in UK), in future larger units	No	Yes	Yes
Wind	Up to 5 MW	Yes	Not accurately	No
Landfill gas	1 MW	No	Yes	Yes
Run-of-river hydro	kW (in UK)	Yes	Not accurately	No
Photovoltaic cell	1 kW, domestic up to 100 kW commercial	Yes	Not accurately	No
Wave	No commercial examples yet	Yes	Not accurately over long term	No
Tidal	No recent	Yes	Yes	No

[a] Whether output varies in time.

2.10 Combining Sources

Many power systems at present make use of renewable sources, most commonly hydro and wind power. For now it is sufficient to note that, as mentioned earlier, geographical diversity of the wind resources assists integration. Geographical diversity also applies to the other renewable resources and is important to their integration into power systems. The relationship between the different renewable sources, spatially and temporally, is also directly relevant to the issue of integration. Tidal current velocities are not correlated with wind speeds, and wave energy, although an integrated form of wind energy, is not strongly correlated with ten minute or even hourly wind speeds. Solar power has its own unique pattern of variation, and biomass is flexible. How renewable sources of generation can be combined to meet load demand is a critical issue and is discussed in later parts of the book.

References

[1] Star, F. 'Flexibility of fossil fuel plant in a renewable energy scenario: possible implications for the UK', in *Renewable Electricity and the Grid*, (ed. G. Boyle), Earthscan, 2007.
[2] Moreira, J.R. and Poole, A.D. 'Hydropower and its constraints', in *Renewable Energy*, (ed. T.B. Johansson et al.), Earthscan, 1993.

[3] Bardsley, W.E. and Choudhry, S. 'An approach to estimating hydro power system income gain from computer-ized water scheduling', *Natural Resources Research*, September 2000, **9**(3), 215–222.

[4] 'Wind energy: a vision for Europe in 2030', Report from TPWind Advisory Council, European Wind Energy Technology Platform, 2007.

[5] Conradsen, K. and Nielsen, L.B. 'Review of Weibull statistics for estimation of wind speed distributions', *J. Climate and Applied Meteorology*, August 1984, **23**, 1173–1183.

[6] Seguro, J.V. and Lambert, T.W. 'Modern estimation of the parameters of the Weibull wind speed distribution for wind energy analysis,' *J. Wind Engineering and Industrial Aerodynamics*, 2000, **85**, 75–84.

[7] CREST MSc Wind I notes, Loughborough University, 2007.

[8] Sharpe, D. Chapter 3 in *Wind Energy Handbook* (eds T. Burton, D. Sharpe, N. Jenkins, E. Bossanyi, et al), John Wiley & Sons, Ltd, Chichester, 2001.

[9] Hansen, M.O.L. *Aerodynamics of Wind Turbines*, James & James, London, 2000.

[10] Manwell, J., McGowan, J. and Rogers, A. *Wind Energy Explained*, John Wiley & Sons, Ltd, Chichester, 2001.

[11]Sinden, G. '*Wind power and the UK wind resource*', Environmental Change Institute, University of Oxford, 2005.

[12] Green, M.A. *Solar Cells: Operating Principles, Technology, and System Applications*, Prentice-Hall, Engle-wood Cliffs, New Jersey, 1982.

[13] Nelson, J. *The Physics of Solar Cells*, Imperial College Press, London, 2003.

[14] http://www.eere.energy.gov/consumerinfo/pdfs/solar_overview.pdf.

[15] Farrell, A., et al. 'Ethanol can contribute to energy and environmental goals', *Science*, January 2006, **311**, 27.

[16] UK DTI report URN 01/797.

3

Power Balance / Frequency Control

3.1 Introduction

The essential function of an electrical power system is to meet the energy demand of consumers. This chapter describes the power demand characteristics of consumers and the technical and operational principles involved in the reliable delivery of power to them from a variety of conventional and renewable energy generators. The detailed regulatory issues of control and operational requirements adopted in different countries where electricity systems have been liberalized are addressed in Chapter 7.

3.1.1 The Power Balance Issue

Consider first the case in which the power system consists of a single generator driven by a prime mover and supplying a load. In Chapter 4 it will be shown that the frequency of the generated voltage is directly proportional to the rotational speed of the generator.

The generator has rotational inertia and it is assumed that the prime mover is fitted with a *governor*. The function of a governor is to sense any changes in speed and to adjust the fuel supplied to the prime mover so that the speed (and therefore the frequency) is controlled.

The law of energy conservation requires that at any instant the power demanded by the load, usually referred to as *demand* or just *load*, is supplied by the generator and/or by energy stored within the system. If the load is suddenly increased, the extra energy demand is initially supplied by the rotational inertia of the generator through a decrease of its speed. This decrease in speed is also reflected by a proportionate decrease in frequency. The governor senses this decrease in speed and increases the fuel supply to arrest the fall in speed and frequency. Exactly how far the frequency falls, how quickly it recovers and the frequency of the new equilibrium state depends on the governor characteristics and the frequency dependence of the load. All these issues will be discussed later in more detail. For now it is sufficient to recognize the interdependence between demand and frequency.

Complex power systems consisting of multiple interconnected generators supplying a large number of consumers respond much as the single generator, but with all the generators and loads contributing to the system response during demand changes. Just as for a single generator supplying a load, the frequency of a complex interconnected system is the same at all parts of the network. For now, the idea of a unique system frequency should be accepted without question, but it will be properly explained in Chapter 4. A power system is, of course, never in equilibrium because the demand varies continuously as consumers switch on or off their loads. It can be concluded that frequency shifts are an indicator of the imbalance between supply and demand at a particular instant. Frequency drifts downwards when demand exceeds supply and vice versa.

Conventionally, power systems are run so that their frequency remains within narrow bounds because:

- This ensures that electric motors operate at virtually constant speed. A fixed speed is required in many consumer applications where an AC electric motor is used to drive a device at an approximately constant rate, e.g. a pump in a washing machine or a lathe in an industrial workshop.
- In electronic applications the mains frequency can be used as a basis for timing various processes.
- Transformers are sensitive to frequency variations and may be overloaded if the frequency drifts substantially from the nominal.
- Finally and most importantly, in traditional power stations the performance of the generators is dependent on the performance of all the auxiliary electric motor drives that deliver fuel and air to the boiler, oil to bearings and cooling services to several systems. If these auxiliaries underperform due to low speed caused by low frequency, power station output can be reduced. As will be discussed later this phenomenon could lead to a runaway situation with cascaded shutdown of power stations and blackouts.

For a near-constant frequency to be maintained it is necessary that the supply of power accurately tracks the variations in demand. How and why electricity demand changes is the subject of the next section.

3.2 Electricity Demand

3.2.1 Demand Curves

Figure 3.1 shows the highly variable nature of the electricity demand over a day of a typical individual house in the UK, with a minimum demand of a few watts, an average between 0.5–1 kW and the maximum in the range of 5–10 kW, i.e. 10 to 20 times the average load. Using a dedicated electricity generator to supply this house alone would be hopelessly expensive. The generator system would have to be large enough to meet the maximum demand, but most of the time it would be running at a very small fraction of its rated capability. Any fuel driven prime mover operated in this way would be woefully inefficient. Energy storage,

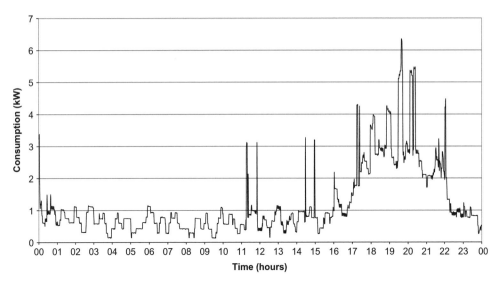

Figure 3.1 Demand curve of an individual house

such as batteries, could improve generation efficiency, but would have to be quite large, and therefore expensive, in relation to the average demand. If each household in the UK met its own maximum demand of say 5 kW, 100 GW of plant would be needed for this sector alone.

3.2.2 Aggregation

The smoothing benefit arising from aggregation is of vital importance to electricity utilities. The more uncorrelated the demand among consumers, the more effective the overall smoothing. For a large power system this statistical effect is dramatic and is illustrated by the characteristic demand profile shown in Figure 3.2. This shows a typical demand curve over a day for the whole of England and Wales and should be contrasted with Figure 3.1. Not only is the peak to mean ratio considerably reduced but this curve is noticeably much smoother than that of the individual house. As a consequence it is much easier to predict, and the generation required to supply this aggregate load can be scheduled and controlled very efficiently, as will be discussed later.

The value of interconnection to form large power systems should now be clear: it allows demand aggregation and the benefits that stem from this, primarily through the easier matching of supply and demand. Some proponents of renewable energy suggest that national grids will become redundant once generators are located near to consumers, but this is a misconception unless an unprecedented breakthrough in energy storage technology is achieved. Indeed, given the intrinsic variability of many dispersed renewable energy sources, interconnection may well prove to be even more valuable in the future.

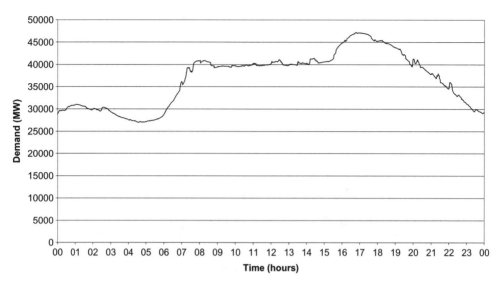

Figure 3.2   Illustration of the smoothing effects of aggregation. (Source: National Grid plc)

3.2.3 Demand-side Management – Deferrable Loads

Two important linked concepts are those of *deferrable loads* and *demand-side management*. A deferrable load consumes a certain amount of energy to provide a service but is flexible in terms of exactly when that energy is supplied because it possesses either an internal storage capacity or a large thermal inertia or because the consumer is flexible about the time when he or she requires the energy service.

Demand-side management is a technique used by utilities to regulate remotely the demand required by deferrable loads so that their connection to the grid is scheduled according to the availability or cost of power. The use of off-peak electricity tariffs for night-storage heating is a familiar example in the context of the UK and helps to level out the demand difference between night and day so that a greater proportion of the generation can run at a constant output. Similar incentives are offered to commercial and industrial users, but, on the whole, consumers connected to the large reliable power systems found in affluent countries rarely give much thought to *when* they use electricity. For the most part, the generation in these systems is designed and controlled to meet the load, rather than the other way round.

Conversely, users of standalone electricity systems often time their use of electricity to suit the available generation. In fossil fuelled systems, it is usually a matter of levelling out the load. In systems relying mainly on wind or solar power, the timing of use is obviously related to the availability of the resource.

Exploiting the deferability of loads is a useful tactic in any power system; it is especially valuable in systems relying on variable renewable energy sources, and can be far cheaper than employing energy storage.

3.3 Power Governing

3.3.1 Power Conversion Chain

To understand the impact of RE sources on power systems it is essential to understand the way in which thermally driven generators are controlled. This is also of direct application to biomass fuelled systems, where in many cases the plant is similar to a standard fossil fuelled plant.

Figure 3.3 shows the power flow diagram for a simple fuel fired generator [1], supplying electricity to consumers as depicted in Figure 1.12 in Chapter 1. Energy is first converted from chemical into thermal form in the boiler, from thermal into mechanical form in the turbine, from mechanical into electrical form in the generator and finally back to thermal, mechanical, light or chemical form by the action of the consumers. For the frequency to remain fixed the fuel energy input to the boiler must be controlled to balance the variable consumer demand.

Importantly, the chain of power conversion units in Figure 3.3 contains two stages with intrinsic energy storage. The boiler consists of kilometres of piping that carry superheated high pressure steam. Because of its large physical size it constitutes a substantial thermal store, in fact it contains enough energy to supply the turbine at full output for a few minutes. The turbogenerator itself, because of its substantial rotating mass and its high rotational speed, also contains inertial stored energy, in this case only of the order of a few full-output seconds. These stored energies can be used to follow variations in demand over short periods. Rotational energy in the turbogenerator is instantly available in response to decreases in system frequency. A sudden increment in demand will result in a frequency drop that will extract a part of the stored energy from the spinning mass. The more significant stored energy in the boiler is also available, with some delay, to supply any increments of electrical generation through the action of the governor.

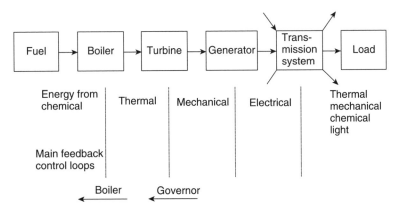

Figure 3.3 Power flow diagram for a thermal power station [1]. (Reproduced with permission of McGraw-Hill)

3.3.2 The Governor

The steam flow from the boiler to the turbine of a large conventional fossil fuelled generator is regulated by a valve of considerable size and weight. For the turbogenerator to respond quickly to a requirement to increase or decrease its output, this valve has to be opened or closed as quickly as possible. This is achieved through a hydraulic actuator. The control signal for the hydraulic actuator is provided by the governor. In the past governors were of mechanical nature but modern generators are fitted with electronic governors. The governor's function is to measure the rotational speed of the generator, to compare it to the reference value (50 or 60 Hz) and, based on the error signal, to instruct the hydraulic actuator to open or close the steam valve.

Figure 3.4 shows the action of the governor of a single generator supplying a load with 50 Hz AC. Governors are designed to operate as proportional control systems; i.e. an error must be present between the set point frequency and the actual frequency for the governor to alter the fuel or steam supply. This proportional control is characterized by the line in Figure 3.4 which has a fall or *droop* of 4% across the operational range. This 4% value has been found through extensive experience to be appropriate for stable governing and is widespread. Governors have the inbuilt facility that allows an operator to adjust the frequency at which the characteristic intercepts the frequency axis, known as the *set point*. For the line aa the set point is 52 Hz. With this set point, the system frequency is 52 Hz when the generator is supplying no load and is reduced incrementally to 50 Hz when the load increases to rated generator power. Changing the set point to 51 Hz moves the line to bb. Set point adjustment is of considerable importance because it allows power system operators to decide how the demand is shared by the generators on the grid, but more will be given on this later.

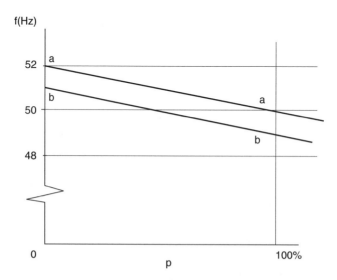

Figure 3.4   Frequency/power governor characteristic

3.3.3 Parallel Operation of Two Generators

When two generators are connected in parallel and are jointly supplying the demand in a small power system, the load is shared according to the set points of their governors. The best way to illustrate this sharing mechanism and the resulting system frequency is by means of an example.

In the small power system of Figure 3.5(a) two generators A and B rated at 50 and 100 MW respectively supply a load of 100 MW. Both generators are fitted with governors having a droop of 4% and a no-load set point of 52 Hz. Lines aa and bb in Figure 3.5(b) show the frequency–power (f–P) characteristics of the two generators.

The division of load between the generators when they are supplying 100 MW will now be determined. Let line cc located Δf below 52 Hz represent the frequency at which this sharing takes place.

(a) Using trigonometry $P_A/\Delta f = 50/2$ and $P_B/\Delta f = 100/2$. Therefore $P_A/P_B = 1/2$, but as $P_A + P_B = 100$ MW, then $P_A = 33.3$ MW and $P_B = 66.6$ MW In this case the demand is shared in proportion to the generator ratings

(b) To determine the system frequency, from $P_A/\Delta f = 50/2$ $\Delta f = 33.3 \times 2/50 = 1.33$ and hence the frequency at which line cc intercepts the f axis is $f = (52 - 1.33) = 50.67$ Hz

(c) Suppose that generator A is more modern and therefore more efficient in terms of fuel used per generated kWh. The set point adjustment of the governors could be used to arrive at a new more economic load-sharing condition that maximizes the contribution from generator A.

One way to implement this is to shift the set point of generator B so that its characteristic is displaced from bb to BB, with the result that each generator is supplying 50 MW. The system frequency is now 50 Hz and the new set point of generator B can be found from $\delta f/50 = 2/100$. Hence $\delta f = 1$ and the new set point is 51 Hz

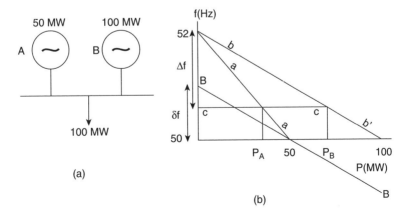

Figure 3.5   Parallel generators. (Reproduced with permission of McGraw-Hill)

3.3.4 Multigenerator System

The two-generator load-sharing paradigm can be extended to explain the behaviour of a typical power system where a very large number of generators are interconnected through a transmission network to feed a multiplicity of loads. If all generators have the same standard governor droop characteristic and a common set point, then they will all share load in proportion to their ratings. In fact, for the economic reasons illustrated in the example of Figure 3.5 and discussed at length in Chapter 7, generators with the lowest cost of energy production are run with their governor set points adjusted at full output (e.g. characteristic bb in Figure 3.5). As such generators are incapable of providing increments in power their governors are unresponsive to demand increases. A number of generators that are less economic to operate, or for other reasons to be explained later, are run with governor set points that results in them operating at part load. Such generators are capable of responding to demand increases. It can be shown that in such well integrated systems and to a good approximation all the generators with *active governors* can be lumped together into one very large generator having a rating equal to the sum of the ratings of the constituent generators and possessing the typical 4% droop due to the combined governor action of all the machines. This *equivalent* generator has a rating several times the value of the typical rating of generators on the network. The reason for this will be explained later. The f–P characteristic of such an equivalent generator is shown by aa in Figure 3.6, where f is the system frequency at which the present steady demand P_d on the equivalent generator is supplied. Because of the large size of this equivalent generator, aa intercepts the P axis far to the right. Suppose that an additional generator is connected to the network where the 4% droop characteristic is given by bb. At the system frequency f the additional generator injects into the system a power ΔP. The power balance principle now demands that the system frequency rises by an amount such that the power injection by the equivalent generator is reduced by ΔP. The figure shows that this requires a very small frequency rise Δf. It can be concluded that in a large interconnected power system, an increase or decrease of power from a single generator will have a small effect on frequency.

A power system that is fed by active governor generation capacity overwhelmingly larger than the rating of a single generator is known as an *infinite bus*. The frequency of such a

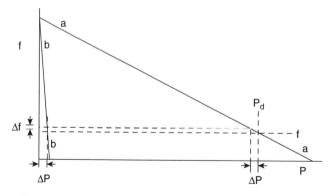

Figure 3.6 Frequency power characteristic of an infinite bus

system is taken to be independent of the power injected by a single generator. In Figure 3.6 this 'independent' frequency is given by f. Therefore, the power injected by each individual generator is defined by the intersection of its characteristic with the frequency line f. By shifting line bb parallel to itself, i.e. adjusting the governor set point, the power injected by each generator can be regulated to the desired level.

3.3.5 The Steady State Power–Frequency Relationship [1]

The new frequency at which a power system will stabilize after a sudden change in generation or demand depends on:

(a) the combined effect of the droops of all active governors;
(b) the total load on the system and its frequency dependence.

The small-increment frequency–power behaviour of a power system can be described approximately through a simple mathematical relationship.

(a) When the frequency changes by $-\Delta f$ (a frequency drop due, for example, to a sudden demand increase), governor action increases the power output of generators by, say, ΔP_G.
(b) Consumer demand consists of a variety of loads. All resistive loads i.e. heating and lighting as well as electronic devices such as PCs, are, in general, insensitive to small frequency variations. However, the large majority of electrical drives, whether in household appliances or most importantly in industry, are frequency sensitive unless they are connected to the mains through a power–electronic interface. A large induction motor drive will run marginally slower at a lower frequency, provide less mechanical power and therefore absorb lower electrical power. The decrease in total system demand can be defined by $-\Delta P_D$ for a drop in frequency $-\Delta f$.

It can now seen be that for a drop in frequency of $-\Delta f$, the effective increase of available power ΔP is partly due to the increase of power from generators and partly due to the *released demand* from loads. Therefore,

$$\Delta P = \Delta P_G + \Delta P_D \quad \text{for a frequency change of } -\Delta f.$$

If it is assumed that the small variation ΔP is linearly related to the frequency change Δf, the power frequency relationship is therefore given by

$$\Delta P = -K\Delta f \tag{3.1}$$

where K is a constant expressed in MW/Hz and is known as the *power–frequency*, or *frequency response characteristic*, or *system gain*. The negative sign in the equation indicates that a drop in frequency results in an increase in the net power available within a system. This relationship gives a rough value of the resulting system frequency after the initial transients have subsided in the case of a sudden major loss of generation or less usually of a sudden increase in demand. The relationship describes the situation maybe a minute after the incident (see Figure 3.10, discussed later) therefore K is not related to rotating inertia effects. The value of K for a specific power system depends on the governing characteristics of the generators and the combined frequency–power sensitivity of the loads. Its value varies during the day as the share of load may shift between frequency sensitive and insensitive consumers.

The value of K can be estimated by tripping out a large generator and measuring the resulting frequency drop after the transients have subsided. Preferably, values of K can be derived at random times from the records of system response after the loss of a generator due to a fault. To ensure stable operation after the most severe contingency that the utility considers as *credible*, i.e. probable, K should be such that the system frequency does not drop below a value that leads to serious underperformance of power station auxiliary drives. Of course, the utilities cannot protect the system from contingencies that have very low probability, e.g. the simultaneous loss of two large power stations. If such events were to occur, measures are in place to maintain the integrity of the system by means of unavoidable power interruption to some consumers, but more of this will be given later.

3.4 Dynamic Frequency Control of Large Systems

3.4.1 Demand Matching

Figure 3.2 showed the daily variation of the aggregate load for a typical utility. The pattern of low load at night and the peaks in the morning and afternoon is a feature of such demand curves. Superimposed on this daily cycle are faster random variations.

Power systems operation is conventionally broken down into different timescales ranging from seconds to days. During second to minute load variations partly loaded plants respond through governor action. Generating plants responding in this timescale are said to be providing *continuous service* or *frequency response*.

The next time scale is *load following*, which involves the connection/disconnection of plant to balance the anticipated load increases/decreases. This timescale covers approximately 10 minutes to several hours during which decisions are taken in response to the trend in demand on the basis of plant operation economics. For example, in anticipation of the early morning increase in demand, the system operator either directly or through bidding mechanisms (Chapter 7) is responsible for ensuring that sufficient capacity is available to meet these large but relatively slow and predictable changes in demand.

Fossil fuelled generators, and especially plant fuelled by coal, can be readily made to follow demand. Such generators are brought on line to supply any escalating demand according to cost effectiveness and flexibility. This results in the layering of generation shown in Figure 3.7. Nuclear power being inflexible occupies the bottom of the pile. CCGT generation and efficient coal occupy the next layer. These two layers supply what is known as the *base load* with other coal fired generation and imports from adjacent networks required to do the load following. Pumped hydro, when availble, is used to extract energy when demand is low and to deliver it during peak demand.

For load following to be effective it is essential to have accurate predictions of the demand curve. Programming generation at this timescale is known as *unit commitment* or *generation scheduling*.

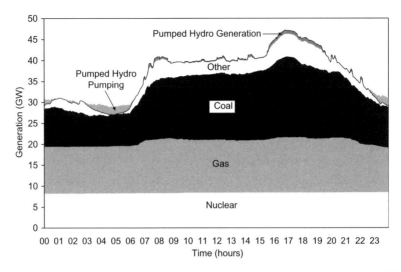

Figure 3.7   Demand curve of the England and Wales power system (Source: National Grid plc)

The reader should not misunderstand the nature of the layering in Figure 3.7. For example, it would be wrong to assume that a constant load like office lighting is supplied from a nuclear power station while a toaster is supplied from coal! Rather a power system could be likened to a bathtub in which water is fed from all the generators and is extracted by loads. The level of water in the bathtub would represent the frequency of the system which the operators endeavour to keep constant. Some generators, e.g. nuclear, provide a constant inflow of water. Other generators are instructed to supply water when the level is detected to be falling. In such an analogy there is no way someone can tell which 'generated water molecule' reaches which load. In other words, electrons cannot be labelled.

3.4.2 Demand Forecasting

Accurate forecasts of demand are required because:

- Electrical energy cannot yet be stored economically.
- The largest proportion of generating plant is thermal in nature. An unfortunate feature of this plant is the considerable delays involved in preparing the cold generators for connection to the power system (several hours) and the restrictions in the rate at which a steam driven turbogenerator can be loaded after connection. These operational delays are dictated by the thermal/mechanical safety requirements of massive boilers and of turbogenerator sets.
- Thermal generators using steam turbines have an upper limit of power generation equal to their *nameplate rating*, but also a lower limit dictated by cavitation problems in the turbine blades at low throughputs of steam. Consequently, when a turbogenerator is connected to the network it should be loaded to a level at least equal to the minimum recommended by the manufacturers (from 30 to 50% of rated power).

Figure 3.7 shows that there are periods during the day (e.g. 6 to 7 am) when the rate of demand growth is considerable. To maintain system frequency, the injected power must closely track the trajectory of the demand curve. Unfortunately, because of the sluggishness of the thermal plant, this tracking cannot be done unless preparative action is taken some hours before the event.

It may be concluded that there is an absolute necessity to carry out a *demand forecasting* activity in order to prepare and progressively load plant as required. Utilities have invested considerable effort in forecasting the daily pattern of demand. Through years of experience they have evolved sophisticated mathematical techniques to correlate demand to the aggregate of the national habits and to other factors such as weather. All methods are essentially based on the fact that demand exhibits regular patterns. Forecasting techniques adjust past demand to present weather and other conditions. Meteorological data on temperature, wind speed, humidity, cloud cover and visibility are used as variables because such factors have an important bearing on heating and lighting demand. The art of load forecasting has been refined to such an extent that estimates are rarely in error by more than ±3% and on average in the UK system they are accurate to within ±1.3%.

Demand prediction techniques are constantly being refined but there will always be occasions when unforeseen circumstances increase or depress the load. The average daily errors in demand in a typical month on the English system are shown in Figure 3.8; during this period the maximum error in prediction was just under 4% and on average it was less than 1.53%. The standard error during this month was 1.6% which, as the average demand was about 32 GW, corresponds to about 300 MW. The figure shows that on 11 November the forecast was adrift by 3.5%, representing a maximum error of over 1 GW.

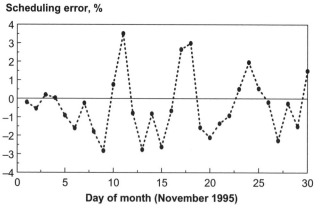

Source: Electricity Pool
Standard deviation: 1.6%

Figure 3.8 Typical scheduling errors on the network in England and Wales. (Reproduced from Milborrow, D., 'Wind power on the grid', in: Boyle, G. (ed.), Renewable electricity and the grid – the challenge of variability, with permission of Earthscan, 2007)

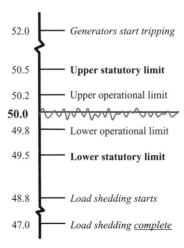

Figure 3.9 Operational and statutory limits of system frequency for the UK power network. (Source: National Grid plc)

In conclusion, even if all generation were 100% reliable, a substantial reserve would still be required because it is simply not possible to predict the demand on a power system exactly.

3.4.3 Frequency Limits

Although perfect balance between supply and demand is not achievable, the balance has to be good enough so that under normal operating conditions the frequency does not drift outside prescribed operational limits.

Figure 3.9 shows that for the UK, statutory limits on permissible frequency variation are ±1% (95% of week) with absolute limits in emergencies of +4% and −6%.

However, the National Grid's usual operational limits are ±0.4% which are tighter than the legal requirements. Under unusual circumstances e.g. substantial loss of demand, the frequency may rise uncontrollably and generating sets are fitted with overfrequency protection that will trip them off the system if the frequency reaches 52 Hz. At the other extreme, due to loss of generation or less frequently, unexpected demand increase, the frequency may fall below its statutory limit.

Large groups of noncritical loads (i.e. excluding hospitals, emergency services etc.) are fitted with staggered under-frequency protection so that switching off such loads, i.e. load shedding begins at about 48.8 Hz and is completed by 47 Hz. Further drops in frequency could trigger cascade tripping of power stations. Such cascade tripping incidents are not unknown, several having taken place in developed countries during the last decade. Complete shutdown of the electricity supply or blackout is an occasional, but traumatic event for millions of affected citizens trapped in lifts, underground trains, etc. Full restoration of supply may take as long as several days.

The topic of frequency control is of particular importance in systems with substantial renewable energy penetration.

3.4.4 Generation Scheduling and Reserve

Demand forecasting provides a fairly accurate picture of the expected load over the following 24 hours so that enough generators are scheduled to provide the expected demand plus a *reserve*, a concept to be discussed in the following section. The process is complicated by the disparate characteristics of plant on the system.

In Chapter 7, it will be shown how generators are loaded on economic grounds subject to various constraints. In anticipation of an increase in demand a choice has to be made by a utility on which uncommitted generating sets from a number of available generating units (some sets may not be available as they are undergoing repairs or maintenance) have to be prepared (heated, synchronized and loaded) and eventually shut down when the demand declines. This is a complicated economic and technical choice. In a privatized energy market one company, usually the one in charge of the transmission system, has the role of plant selection, but here decisions are taken in the context of the contractual relations between the different participants in the electricity market. Chapter 7 describes how regulatory tools, tariffs and bidding systems are used in these circumstances to ensure that supply tracks demand.

3.4.5 Frequency Control at Different Timescales [2, 3]

The maintenance of frequency involves a *response* from generators over several timescales ranging from seconds to days. This response is not only required to follow the demand's fast variability from second to second and its slow variability over the day, but also sudden substantial mismatches between generation and demand, for example during system faults.

Figure 3.10 illustrates a typical system frequency trajectory plotted on a nonlinear time scale. From the origin of the graph until about 8 seconds, the frequency exhibits the usual noise associated with minor mismatch between the continuously varying demand and the efforts of generation to match it through governor action. At 8 seconds something unpredicted and serious takes place. It could be that a large power station trips because the overhead line connecting it to the transmission system suffers a mechanical failure due to high winds and accumulated snow. This *contingency* results in an instantaneous large shortfall of generation. The trace describes a typical time history of the frequency and the measures taken to constrain the frequency excursion within the statutory and operational limits. Such measures are taken by all utilities but are given a variety of labels. Here the labels shown in Figure 3.10 are the ones adopted in the UK.

A power system has at its disposal a number of generators with diverse characteristics. These are arranged in a hierarchy of plant appropriate for operation at different timescales as described below.

A *continuous* or *frequency response* is provided by generators equipped with appropriate governing systems that control their outputs to counteract the frequency fluctuations that arise from relatively modest changes in demand or generation [2]. Large generators on the grid

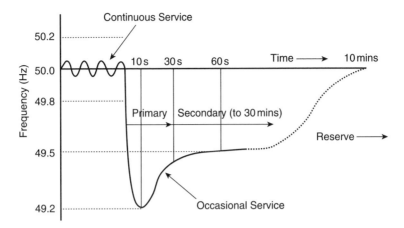

Figure 3.10   Frequency response requirements. (Source: National Grid plc)

are selected on technical and economic merits and instructed by the system operator to operate in frequency-sensitive mode (i.e. under active governor control) to provide this service. For this to be achieved, some generators are held below maximum output.

An *occasional service* or *reserve* is available to contain significant and abnormal frequency excursions caused by sudden mismatches in the generation/demand balance (e.g. loss of generation) [2]. Part-loaded large synchronized generators as well as deferrable loads fitted with frequency sensitive relays provide such services. The reserves are subdivided into *primary* and *secondary* categories. In Figure 3.10 the initial rate at which the frequency drops after the incident at 8 seconds is controlled and limited instantly by the inertial energy release from all the decelerating generators (and consumer drives) on the system. This provides a breathing time before the fast governors of some generators begin to act. Massive steam valves have to be opened hydraulically and increased steam flow has to be transported from the boiler to the turbines. It therefore takes a finite time for the substantial stored energy in the boiler to be exploited.

Primary reserves require the most rapid generator response. The key requirement for generators allocated this task is that they should be capable of increasing their active power output within 10 seconds of predefined system frequency excursions and be capable of maintaining this response for a further 20 seconds [2].

Secondary reserves require a slower initial response but maintained for longer periods of time. This requires the capability of increasing the active power output within 30 seconds and maintaining the response for a further 30 minutes [2]. A fast response capacity is also provided by partly loaded hydro or pumped storage (when available), which are not bedevilled by the constraints imposed on thermal plant. Water driven plant can respond in a few minutes and be started up automatically when the frequency falls below a critical value.

A fundamental feature of generators providing frequency response and reserves, collectively known as the *operating margin*, is the requirement for generators to have *headroom* in order to increase output. The operating margin is the difference between available generation and actual demand. Generators providing such services will therefore be part-loaded.

Besides hydro and pumped storage plant (if available) the main providers of such services are flexible large coal fired power plant.

Beyond primary and secondary reserves, power systems have further robust *tertiary* defences known as *standing* reserves. These are sourced from unsynchronized standby generators capable of mains connection and generation of the instructed level of output within 20 minutes [3]. Typically, standing reserve is provided from generators driven by open cycle gas turbines and reciprocating internal combustion engines.

High frequency response services are required in the event of excessive system frequency events when large loads are suddenly lost. Such services happen rarely and are initiated through governor action requiring either to reduce output or to cease generation altogether.

Finally, utilities have some control over the demand by implementing voltage control. System voltage, like system frequency is rarely exactly at its nominal value but is allowed to vary within controlled limits. One response to a loss of generation, which may occur due to manual intervention or automatically, is a reduction of system voltage. Total system load will fall with voltage depending on the nature of the loads. Voltage and frequency reductions can cope with serious credible demand or generation changes although only in exceptional circumstances would both be allowed to reach their minimum or maximum values.

3.4.6 Meeting Demand and Ensuring Reliability

In the early power systems, a single generator provided power to a local network that serviced a number of loads. In such a system the generator would have to be exceedingly flexible and capable of supplying the lowest to the highest demand. In the case of failure of this generator the supply would be interrupted unless a similar generator was kept in reserve providing a 100% back-up. By interconnecting such isolated systems and allowing exchange of power, the risk of generator failure can be shared so that lower back-up capacity is required to provide the same level of system reliability.

As mentioned earlier, system balancing requires a number of reserves to deal with unexpected demand fluctuations and loss of generation. These reserves are determined on a statistical basis that takes into account the margin of error in demand prediction and the probability of system failures. The aim is to meet specific levels of reliability that historically have proved to be acceptable to customers. To maintain near 100% reliability would be prohibitively expensive. Critical loads such as hospital operation theatres and major computing networks are further backed up by *uninterruptible power supply* (UPS) systems based on batteries or local engine generators.

To maintain frequency stability the *operating margin* or *balancing reserve* provided must be sized in relation to three factors:

1. The maximum deficit that may suddenly occur in a power system due to loss of generation.
2. The expected availability of all plant connected to the system.
3. The demand prediction error.

In the UK system the transmission system operator (TSO) currently carries a balancing reserve of 2.5 GW [4]. Within limits, the larger the available headroom of the frequency response plant the greater the reliability of the power system; however, the presence of these additional part-loaded generating sets on the system significantly increases the system fuel costs. Any part-loaded machine has a poor efficiency due to *standing losses*: fixed losses incurred by a generator when running unloaded.

The risk of demand being unmet is measured by a statistical quantity known as the *loss-of-load probability* (LOLP). In the UK pre-privatized system a LOLP of 9% was considered acceptable, i.e. expected loss of supply incidents for each customer in nine winters per century. The LOLP gives no information on how much load will be shed or for how long. It is understood that reliabilities of the same order are expected in present privatized systems of developed countries although figures are difficult to obtain. For such a reliability to be maintained in addition to the short term reserves discussed earlier, a *system* or *plant margin* must be maintained. This is the difference between *installed capacity*, including imports and exports, and maximum annual *peak demand*. This margin is necessary because at any time there are bound to be plants that have broken down or are out for maintenance. For reliable operation, a power system must have a plant margin that is substantially greater than the operating margin. Taking as an example a system with a 70 GW winter peak demand, to guarantee that the demand will be met to a typical reliability expected in an industrialized nation about 84 GW of conventional capacity (20% above peak demand) should be available on the network.

In privatized or *deregulated* power systems all these activities are managed through the complex financial instruments that define the contributions and obligations of several independent generation providers all feeding into the same network. For example, in the UK and in other similarly privatized networks, system operators pay considerable premiums to power generators who provide response and reserves.

3.4.7 Capacity Factor and Capacity Credit [4]

When renewables displace significant amounts of conventional generation plant, an extra conventional plant capacity margin is required to maintain system supply reliability. It is important to recognize that an additional plant margin will only be required at times of low electricity demand and high input from variable sources. At other times an additional margin will not be required as there will be sufficient conventional plant available to meet demand irrespective of the contribution from variable sources.

The capacity factor of a generator, as mentioned in Section 2.4.3, is usually defined as the ratio of its yearly energy output to the output it would have produced if it operated continuously at its nameplate rating. Due to unavailability caused by maintenance schedules, breakdowns, etc., no plant achieves a capacity factor of unity. Base load thermal generators have a capacity factor of 85–90% when new, declining over the years until they are decommissioned. Wind turbines achieve capacity factors of 20–40% depending on the windiness of the site. In the UK the average capacity factor of onshore wind farms is ~30% with offshore schemes achieving higher values. Often, uninformed commentators quote the figure of 30% when they claim that wind turbines require back-up for 70% of the time. In fact wind turbines provide some power between cut-in and cut-out wind speeds for 80% of the time. The capacity factor does not determine back-up requirements, which must be assessed statistically.

Statistical analysis confirms that, for limited penetration, the capacity factor is a good guide of the probability that the generator in question will be available to contribute towards meeting demand [5]. On this basis variable sources are incapable of providing the same level of reliable or *firm* power during demand peaks as conventional generators, but they are still capable of providing a contribution to this. This firmness is known as the *capacity credit* and is a measure of the amount of load that can be provided by variable plant with no change in the LOLP. This can be illustrated by an example. A 1 GW dispersed installation of wind farms with a 30% load factor will provide the same yearly energy output as a 350 MW gas fuelled plant of 85% load factor. Both plant statistically have a capacity credit of 300 MW.

The calculations above assume that the availability of generated power is completely uncorrelated to the demand. Although this assumption is correct for the gas fuelled plant this is certainly not so in the case of the wind plant. As will be shown in the next section, due to aggregation of wind farms and the characteristics of the wind resource, there is a positive correlation between the availability of wind power and the demand.

3.5 Impact of Renewable Generation on Frequency Control and Reliability

3.5.1 Introduction

The introduction of variable RE generation into a network will have an impact and incur associated costs in two main categories [4]. The first can be labelled as the *balancing impact* and relates to the management of demand fluctuations from seconds to hours. The second, referred to as the *reliability impact* relates to the requirement that there is enough generation to meet the peak demand. Both balancing and reliability involve statistical calculations. The introduction of variable RE generation introduces additional uncertainties that can be quantified in terms of operational penalties that have to be taken into account when the value of electricity from RE sources is calculated.

There is a widespread, but mistaken, belief that operation of an electricity system with renewables causes serious problems. A common misconception is that significant additional plant must be held in readiness, to come on-line when the output from the wind plant ceases. This would indeed be true in an island situation, with, for example, wind the principal source of supply. Modest amounts of variable renewables within an integrated electricity system pose, however, no threat whatsoever to system operation. The reason for this is that these amounts do not add significantly to the uncertainties in predicting the generation to ensure a balance between supply and demand. Therefore the risk of changes in the output from variable renewable sources has only a small influence on the needs for reserves.

In the following sections the discussion will be limited mainly to the integration issues of wind power because this is currently the nonschedulable renewable energy source making the largest impact, and is likely to remain so for the foreseeable future. In a later section the impact from other renewables will also be briefly considered. The discussion will review

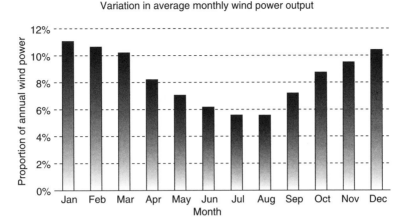

Figure 3.11   Variation in monthly wind power output. (Reproduced from Sinden, G.E., 2007, D Phil Thesis with permission of Environmental Change Institute, Oxford University Centre for the Environment)

the temporal availability and operational aspects of wind power and the penalties incurred for different penetration levels so that the system reliability is maintained at the desired level.

3.5.2 Aggregation of Sources

In Section 3.2.2 it was shown how integrated electricity systems benefit immeasurably from the aggregation of consumer demand. Fluctuating sources can benefit in the same way. Figure 2.9 in chapter 2 confirms the benefits of adding the output of geographically dispersed wind sites in the UK [6]. Aggregation here has provided its usual benefits by smoothing the output over short and medium timescales.

The Monthly Distribution of Energy

The seasonal wind power availability from dispersed sites in the UK shown in Figure 3.11 indicates limited production during summer and greater than average production during winter [6]. On average twice as much electricity is generated during the winter compared to the summer months. This pattern matches the seasonal demand pattern in the UK.

The Daily Distribution of Energy

Figure 3.12 shows that, on average, wind power availability is higher during the daytime than at night in the UK [6]. This trend is present irrespective of the time of year and is of benefit in a system where the demand peaks during the afternoon period when the wind power availability is near its maximum.

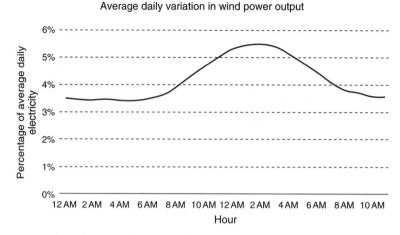

Figure 3.12   Average daily variation in wind power output. (Reproduced from Sinden, G.E., 2007, D Phil Thesis with permission of Environmental Change Institute, Oxford University Centre for the Environment)

Short Term Variability

The variability of wind power will cause changes in the power generated from one hour to the next. The maximum expected rate of change from hour to hour provides an indication of the reserves required to deal with shortfalls in supplying demand. Wind speed variations within the 15–25 m/s band will result in no change of power as the wind turbine will be operating at full output for winds in this range. However variations within the band 4–15 m/s will result in substantial power changes. The degree of dispersion of the resource will again be of advantage as increments of wind at one site will be compensated by decrements at another.

Figure 3.13 shows the benefit of geographical dispersion from a German study [7]. In the graph, the frequency of occurrence of positive and negative changes observed in the hourly values of wind power output from actual wind farm measurements is shown for a single wind farm and for wind power aggregated over the whole of Germany. Whereas a single wind farm can exhibit hour to hour power swings of up to 60% of installed capacity this figure is less than 20% for aggregated wind farms. These maximum changes are likely to occur about once a year. These figures are significant because they indicate the requirement of fast response part loaded thermal plant to *ramp-up* or *ramp-down* and this would have economic implications as discussed later.

The Capacity Factor

Figure 3.14 shows the results of a UK study on the relative capacity credit between seasons and between onshore and offshore resources as a function of wind power penetration [7, 8]. Due to the windiness of the UK offshore areas the credit is higher than 60% during winter, decreasing to less than 25% in the summer for small penetrations. The onshore picture is

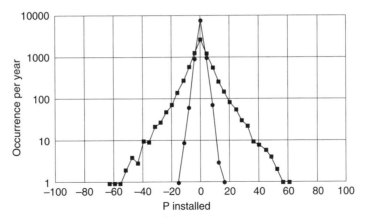

● Wind power aggregated over Germany
■ Single wind farm, inland Germany

Figure 3.13   Frequency of occurrence of hourly wind power variations. (Reproduced from Reference [7], EWEA Report, 2005)

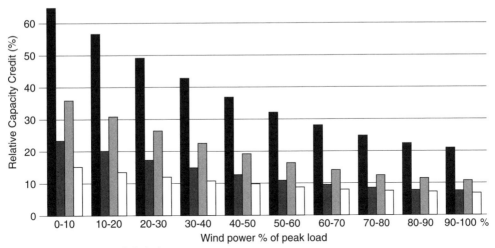

■ Offshore capacity credit (%) winter
■ Offshore capacity credit (%) summer
▨ Onshore capacity credit (%) winter
☐ Onshore capacity credit (%) summer

Figure 3.14   Seasonal capacity credit as a function of penetration for onshore and offshore UK wind resource. (Reproduced from Reference [7], EWEA Report, 2005)

similar but scaled down. As expected, the capacity credit decreases with increasing penetration. It may be concluded that, in the UK, because of the favourable weather conditions, a statistical correlation exists between the availability of wind power and demand. The capacity credit associated with wind power and therefore the contribution of wind towards reliability could be substantially higher than that calculated from the annual average wind capacity factor.

3.5.3 Value of Energy from the Wind

The financial benefit of wind power can be calculated by determining the cost of supplying the total demand and then subtracting the cost of supplying the residual demand (after wind power is added) from the existing or a reduced mix of generation. In practice this calculation is extremely difficult to do and so, for simplicity, many analysts concentrate on fuel savings which are more straightforward to estimate. *Ideal fuel savings* are simply those calculated from the cost of displaced generation and do not take account of any changes of operation forced on conventional plant by the time varying characteristic of the wind plant. To estimate the fuel savings more realistically the following operating issues should be considered [9].

- If wind generation is subtracted from the gross demand (wind power can be viewed as negative demand) the residual demand will have more variability than the gross demand. This means that the output of fossil fuel plant providing a continuous or frequency response service will have to be adjusted more frequently and to a greater extent. Additionally, with wind generation there is a need to ensure that the system can respond adequately to unpredicted changes over longer time periods. Extra balancing reserves provide greater headroom, but the lower loading level of thermal plant that this requires would result in lower operational efficiency of thermal plant. Finally, at higher penetrations (above 20%), some thermal plant may have to be shut down and started up to maintain adequate reserves. This will incur what are known as cycling costs. All of the above effects make up the *balancing impact*.
- Wind generated electricity may increase the size of the system margin required to maintain the required level of reliability. The reason for this is that wind plants are less likely than thermal plants to be available to contribute towards times of peak demand. This effect makes up the *reliability impact*.
- At high penetrations, energy may need to be *spilled, discarded* or *curtailed* because for operational reasons this energy cannot be safely absorbed while maintaining adequate reserves.

The issues that define these impacts and the associated costs and penalties that have to be assigned to wind generated electricity are discussed in what follows.

3.5.4 Impact on Balancing

Thermal plant incur some costs when they are run in the frequency control mode. It can be shown, [5, 9–11] that if variations in a variable source and in demand occur roughly independently, the total resulting variation in the net load to be met by the thermal plant is approximately a 'sum-of-squares' addition of the components:

$$\begin{pmatrix} \text{Total variability} \\ \text{of load on thermal} \\ \text{units} \end{pmatrix}^2 = \begin{pmatrix} \text{Total variability} \\ \text{of electricity} \\ \text{demand} \end{pmatrix}^2 + \begin{pmatrix} \text{Total variability} \\ \text{of variable} \\ \text{source} \end{pmatrix}^2 \qquad (3.2)$$

Thus, for example, when the average power variation of the added source equals that of the demand itself, the total variability is not doubled but increased by 40%. This has some important implications. The impact of fluctuations in variable sources at low penetrations can be taken to be practically zero; in other words this impact is just noise added to demand fluctuations.

Over longer timescales, the level of operating reserve required at any given time depends on two key factors: uncertainties in demand prediction and the probability of loss of the largest generation plant on the network. When wind power plant is introduced into the system, an additional source of variation is added to the already variable nature of demand. To analyse the additional variation caused by the wind plant it is important to appreciate that the requirement is that the entire system must be balanced instead of balancing each individual load or resource. The operator has to ensure that the average system reliability is maintained at the same level it would have been without the wind resource.

However, the crucial question is by how much does wind generation increase the balancing uncertainties? Intuitively it is known that minute-to-minute fluctuations in wind output are largely uncorrelated to load. This implies that the additional uncertainty introduced by wind power does not add *linearly* to the uncertainty of predicting the load. As for the issue of variability dealt with above, it can be shown that when errors in predicting the output from variable sources occurs independently of those in predicting demand, the combined error is again a sum-of-squares addition [9–12]:

$$\begin{pmatrix} \text{Average error in} \\ \text{predicting net load} \\ \text{on thermal units} \end{pmatrix}^2 = \begin{pmatrix} \text{Average error in} \\ \text{predicting electricity} \\ \text{demand} \end{pmatrix}^2 + \begin{pmatrix} \text{Average error in} \\ \text{predicting variable} \\ \text{input} \end{pmatrix}^2 \qquad (3.3)$$

Demand prediction techniques are constantly being refined but there will always be occasions when unforeseen circumstances push up or depress the load. Equation (3.3) indicates that for small penetrations of variable sources the prediction errors are lost among load fluctuations. However, since demand is fairly predictable, forecasting errors from substantial penetration of wind will incur some penalty.

Analysis of the combined uncertainties of wind, demand and conventional generation based on the sum-of-squares calculation of Equation (3.2) make use of the standard error in predicting the generation/demand balance. On typical developed country networks, one hour ahead, this averages at around 1% of the demand. For four hours ahead, this figure rises to 3%.

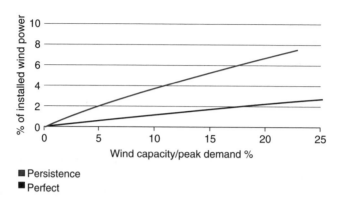

Figure 3.15 Additional balancing power for wind (Published in *Windpower Monthly News Magazine*, December 2003 [12])

The need to schedule reserve to cover for possible trips of conventional thermal plant emphazises the point that no generation is 100% reliable. While the loss of a typical 1000 MW of thermal plant is a real risk, it is almost inconceivable that 1000 MW of wind plant would be suddenly lost. It is also assumed that dispersed wind plant is not sensitive to a common mode disturbance; i.e. the plant rides through voltage dips caused by faults on the transmission system. The more wind that is installed, the more widely it is spread, and sudden changes of wind output across a whole country simply will not in practice occur [13].

Calculations can be made over various timescales to determine the need for extra reserve. Figure 3.15 shows the estimated additional balancing power needed (expressed as a percentage of installed wind power) as a function of wind power penetration [12]. At 20% penetration, 7% of extra operating reserves are required if persistence[1] in the wind is assumed. For perfect forecasting only 2% of additional capacity would be needed.

Clearly these back-up figures are modest, but what about the associated costs? In the UK, the TNO has estimated that 10% of wind penetration would increase balancing costs by £40 m a year, which is equivalent to £0.002/kWh. A 20% penetration will increase the cost to £0.003/kWh. These figures should be viewed in terms of the retail cost of electricity, which at the time of writing is in the region of £0.10/kWh [14]. Hence a 20% penetration in the UK would incur a 3% additional cost on electricity at present prices.

For relatively little expenditure the predictability of wind could be greatly improved. This could be accomplished partly through the installation of extra weather data monitoring stations (e.g. anemometry towers a few tens of kilometres from major wind farms) and partly through sophisticated computational techniques. Programmes to provide enhanced predictability are being developed in several countries [15].

Figure 3.16 illustrates this by showing a typical one hour wind forecast using sophisticated techniques against actual output for one wind farm over a period of a week. Such techniques could provide considerable cost benefits in operating reserve.

---

[1] 'Persistence' wind forecasting assumes that the wind power output one hour ahead is the same as at time zero. 'Perfect' forecasting as in Figure 3.15 assumes that wind power can be predicted with total precision.

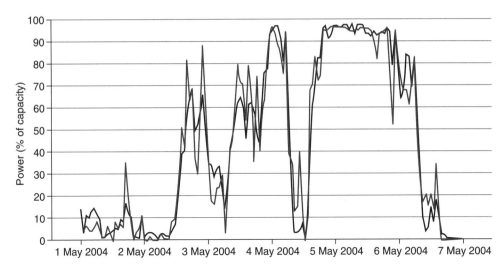

Figure 3.16 Wind farm forecast (+1 hour) versus actual output, Ireland, 2004. (Reproduced with permission of Garrad Hassan and Partners Ltd)

3.5.5 Impact on Reliability [6]

In addition to the operating reserve, some system margin in excess of the system peak demand is required and this will be affected by the level of wind penetration. The example of a power system with a peak demand of 70 GW will be given, [6]. To guarantee long term security to an accepted level of reliability utilities in general would require that about 20% (14 GW) of that peak must be additionally available on the network as the system margin. The system in question has a yearly energy demand of 350 TWh. If 10% of the system energy were to be generated by wind power it is possible to assess what the new plant margin should be. To generate 10% of the energy, i.e. 35 TWh, a wind power capacity of 35 000/(8736 × 0.3) = 13 GW, where 8736 is the number of hours in the year and 0.3 is the average wind farm capacity factor. The 35 TWh would be generated by 35 000/(8736 × 0.75) = 5.3 GW of conventional generation with a yearly load factor of 0.75. Ideally the wind plant would directly substitute for this conventional plant, but the variability of wind power means that not all of this conventional plant can be removed from the network. Statistical calculations by utilities indicate that 81 GW of conventional plant would be required to meet the same level of reliability. Therefore only 3 GW (84 − 81 GW) can be removed from the network; i.e. 2.3 GW (5.3 − 3 GW) is retained because of wind power variability, which is 17% of the wind power installed capacity.

In the UK context of a planned 20% penetration from wind and to maintain system reliability the associated cost lies within 0.03 and 0.05 p/kWh.

3.5.6 Discarded/Curtailed Energy

As the capacity of the variable sources injected into a system increases, there might be occasions when the available power from such sources cannot be used. If the penetration is substantial, there might be periods when the available power from renewables exceeds demand, or cannot be accommodated by the transmission or distribution system. However, even before this stage is reached, energy from variable sources will have to be shed because the power system would need to keep a minimum level of thermal plant generation in order to maintain adequate operating reserve.

Discarding energy from variable sources poses no particular operational difficulties. Output from wind turbines can be controlled through blade pitch variation, from photovoltaics through inverter control and from hydro, wave and tidal schemes through similar control techniques. However, this discarding or *curtailment* of energy results in an economic penalty on variable sources, which becomes increasingly important at high penetrations. This penalty is difficult to assess as it depends heavily on the flexibility of the base load units, i.e. the extent to which they could be operated in a stable regime at low power and upon how rapidly their output could be increased if required. For the level of penetration expected over the next decade or so, the penalties due to discarded energy are unlikely to be of major significance. It should also be noted that curtailment may also be required due to distribution or transmission system constraints. Reference [4] states that 5 out of 6 studies show that at a penetration of 20%, curtailed energy ranges from 0 to 7%. Most studies show that, with sensible design, curtailment due to local network capacity limitations would be rarely required.

3.5.7 Overall Penalties Due to Increasing Penetration

In previous sections, operational penalties due to increases in variable source penetration were reviewed. A number of studies have been carried out to provide estimates of these penalties as the penetration of renewables increases. The majority of these studies relate to wind power, as this is the variable source with the largest installed world capacity to date. Table 3.1 gives some indicative figures based on EU-funded studies and on Danish and UK thresholds linked to operational experience of wind farms. The additional costs are tabulated in terms of the level of penetration.

Table 3.1 Implications of increasing wind energy supply on the UK network

Wind power penetration	Measures required	Cost penalties
Up to 5%	None	Negligible
5–10%	Occasional instances when some energy from the wind is discarded and more part loading of the thermal plant required	0.1–0.2 p/kWh
10–20%	As above, plus more use of pumped storage or hydro to balance wind power	0.2–0.8 p/kWh
20–50%	May be necessary to build more storage, or peaking plant, or retain old coal plant, depending on relative costs (extra storage will benefit the system as a whole)	>0.8 p/kWh

3.5.8 Combining Different Renewable Sources

The benefits of combining different variable sources were mentioned in Section 2.10. As the capacity of a nondispatchable source increases, its marginal value declines, primarily because successive increments of capacity are correlated with those already on the system. In contrast, combining capacity from renewables with uncorrelated or complementary outputs can therefore be of considerable benefit [9, 10, 17].

Typically, a combination of wind and solar could be beneficial. In some circumstances, thermally driven winds can be strongest after sunset, so that the combination of wind and solar usefully covers periods of high demand. Other studies indicate that a combination of wind and tidal (two sources having statistical independence) increases their value compared with the case of having more of the same [9].

A recent study [14] indicates that a contribution from a mix of PV solar and wind plus domestic combined heat and power has the potential to reduce significantly the overall variability that would have been experienced if only one renewable technology were to provide the total contribution.

The potential synergies among different renewable sources are clearly much too important to ignore, and they may often make the combined exploitable potential larger than the sum of the parts considered in isolation.

3.5.9 Differences Between Electricity Systems [4]

It is appropriate here to stress that results from studies on one particular network do not necessarily apply elsewhere. The operational viability and costs of integrating renewable energy depend on a number of factors that characterize the local resource as well as the structure of the electricity network. These factors include:

- the strength and temporal variability of the resource;
- the possibility of geographical dispersion over a large area to gain the advantages of aggregation;
- the possible complementarity between different types of renewable resources;
- the correlation, if any, between availability of the resource and demand variation;
- the extent to which the magnitude of the resources can be forecast, where some weather patterns are more predictable than others;
- the robustness of the electricity network and the proximity of transmission lines to the areas of maximum resource availability;
- the transmission links, if any, to adjacent networks;
- the operating practices of the network, in particular how far in advance the system balancing reserve is planned;
- the type of conventional plant in the network, for example, smaller and more modern thermal plants are more flexible than large base load plant such as nuclear.

3.5.10 Limits of Penetration from Nondispatchable Sources

Early on in the development of renewables, the UK's Central Electricity Generating Board (CEGB) carried out a number of extensive simulation studies to estimate the impact of large

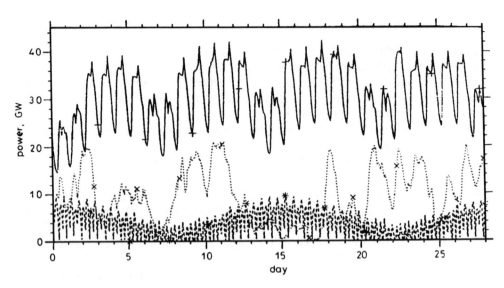

Figure 3.17   CEGB simulation study of large penetration. (From Grubb, M.J., *IEE Proceedings C*, V. **138**(2), March 1991, reproduced with permission of IET)

penetrations [9, 10]. To illustrate the challenges, an extreme example from one such study is reproduced here. Figure 3.17 shows the output that would have been expected from 25 GW of dispersed wind capacity (middle graph) and 10 GW of tidal (lower graph) alongside the demand over a period of one month. In this example 30 and 13% of the consumer energy would be supplied from wind and tidal respectively. The wind power varies less rapidly than demand, tidal more rapidly. Figure 3.18 shows the effect of subtracting the output from wind and tidal power from the demand, leaving a residual load (dashed curve) to be met by a conventional thermal plant.

More recently [18, 19], studies have considered the possibility of meeting the entire demand from a mix of renewables. Figure 3.19 taken from Streater's work [18] shows hour by hour variations in the time variable renewables. Penetration levels of such a high magnitude result in periods when the available power from the RE sources exceeds demand. Even before this stage is reached, for reasons of system reliability, the RE sources would have to be curtailed. Table 3.2 shows the capacities of the different renewables installed and their penetration, defined as the output divided by the total load (i.e. ignoring any curtailment.) It is also apparent that for the year in question a deficit occurs between weeks 15 to 18 and weeks 44 to 52. In principal the shortfall could be made up from biomass based generation.

Denmark and less so, Germany generate a high percentage of their total electricity needs from wind power and are planning further capacity. A recent study from Elkraft, a Danish system operator, asserts that a wind penetration up to 50% is technically and economically feasible. This is based on an increase of installed capacity of 3.1 GW in 2005 to 5 GW in 2025 and a substantial expansion of the Danish grid. The study claims that even with this

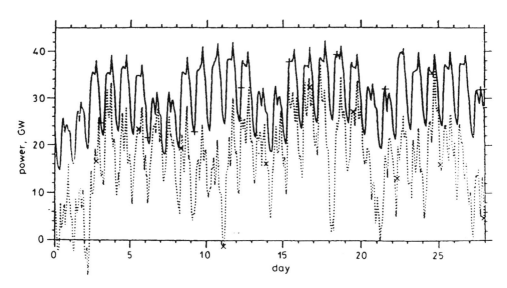

Figure 3.18 Residual demand to be met by conventional generation. (From Grubb, M.J., *IEE Proceedings*, **138**(2), March 1991, reproduced with permission of IET)

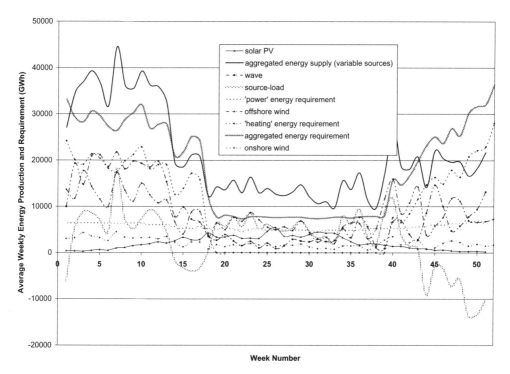

Figure 3.19 Hour by hour variation in renewable energy generation over one year, compared with variations in energy requirements

Table 3.2 Assumed renewable energy capacities and production in the Streater study [18]

	Annual energy produced (TWh)	Percentage of total variable load (%)	Mean power output (GW)
Onshore wind	72 000	20	8.2
Offshore wind	144 000	40	16.4
Solar photovoltaic	36 000	10	4.1
Wave	108 000	30	12.3

penetration, it will be necessary to shut down wind turbines only for a few hours in a year. Similar studies by DENA in Germany [20] indicate that up to 20% penetration by wind is possible with minor extensions of the grid and no need for construction of additional power stations.

3.6 Frequency Response Services from Renewables [2]

With the anticipated rise in the penetration of variable renewables, power systems will be required to accommodate increasing second to second imbalances between generation and demand requiring enhanced frequency control balancing services. Some renewable generation in principle may contribute to frequency regulation services, but this would require headroom in the form of part-loading. Technologies that could potentially provide such services are biomass, water power, photovoltaics and variable speed wind turbines. In Chapter 7 it will be shown that economics dictate that energy from renewable sources should generally be used as fully as possible whenever available. Although this seems to contradict the idea of part loading such plant, there are some occasions when priorities may dictate otherwise.

With large penetrations from renewables there will be occasions, for instance during low demand days over summer, when the number of conventional generators needed to supply the residual load will be so few that an adequate level of response and reserve may be difficult to maintain. Under such conditions renewable generators could be unloaded and instructed to take part in frequency regulation. Such a provision has been made, for example in the Irish (ESB) code, for the connection of wind turbines. In a privatized system the opportunity benefits of running in this mode must more than compensate the loss of revenue from generating at less than the maximum potential.

3.6.1 Wind power

Early wind power technology was mainly based on simple fixed one or two speed stall-regulated wind turbines with little control over the dynamic performance of the generator. However, over recent years active stall and pitch regulated variable speed wind turbines have been developed that are capable of increased conversion efficiencies but also of substantial control capabilities. In principle, modern wind turbines are capable of providing a continuous response by fast increase in power from part loading through blade pitch control in response to drops in frequency and through the same mechanism provide high frequency response through fast reduction in power in response to increases in frequency.

As wind power capacity has increased, to the extent that at times it is the dominant form of generation in parts of Denmark and Northern Germany, there is an increasing demand for wind capacity to be dispatchable and to behave more like conventional generation. Very large wind farms are now expected to conform to connection standards that limit ramp rates for increase in power and also to contribute to frequency regulation under times of high network stress. These requirements are increasingly included in the national *grid codes* that regulate access to the public networks. In these early days it is unclear to what extent this will result in wind power being curtailed, for example to comply with given ramp rates, and to what extent such constraints add value to the system operator.

Conventional steam generation plant assist the network frequency stability at the onset of a sudden imbalance of demand over supply by slowing down. Wind turbines respond differently. The stored energy is in the rotor inertia and fixed speed turbines will provide a limited benefit from their inertia provided that the voltage and frequency remain within their operating limits. Variable-speed wind turbines will not normally provide this benefit as their speed is controlled to maximize the energy production from the prevailing wind.

Large wind turbines are now almost always of the variable speed type and as they increasingly displace conventional generation the total system inertia from such generation will decrease. Consequently the rate of change of frequency and the depth of the frequency dip caused by a sudden loss of generation will both increase. However, variable speed wind turbines could be controlled in principle to provide a proportionately greater inertial energy to the system than conventional plant of the same rating. Such sophisticated control arrangements to support system functions are likely to be requested by utilities as wind penetration increases.

Finally, grid codes require wind turbines to maintain power infeeds to the system even under transient local voltage reductions. Such reductions are usually due to fault conditions in the vicinity of the wind farm. It can be shown that maintenance of power infeed from all generators is essential to ensure system recovery after a fault clearance.

3.6.2 Biofuels [9]

Traditional thermal plant could be described as *capacity limited*, i.e. capable of theoretically generating its rated output continuously, as gas, coal, oil or fissionable material is abundantly available on demand. In contrast, an energy crop based plant could be described as *energy limited* because the locally harvested fuel is limited in nature and may or may not be capable of sustaining all year round continuous plant generation at full capacity. Transporting biomass fuel from remote areas would not be economical.

A biomass plant would be expected to operate as a base-load generator running as far as possible at full output. Such plant would be able to contribute to continuous low or high frequency response services similarly to a conventional plant. For a low frequency response the plant would need to run part-loaded, a convenient strategy providing extra income if, say, due to a low crop yield year the stored fuel would not be capable of servicing continuous full output.

The land filled gas plant size is in the range of 0.5–1.5 MW and because of their small size they would not be suitable for the provision of frequency regulation services.

3.6.3 Water Power

Small and medium sized hydro schemes without significant storage capacity are characterized by substantial variability of output depending on rainfall. Because of this and their small size they are not suitable for frequency regulation duties.

Tidal schemes could be very large indeed and their output would be highly predictable. Such plant would incur exceptionally high upfront capital costs and long payback periods. Operation revenue is vital to service the large loans and it is unlikely that frequency response revenue based on part-loading would be attractive enough. As such schemes are not yet in the planning stages, the jury is still out on their frequency control capabilities.

The comments made above on wind power generally apply with reservations to future wave power schemes. As the technology is still in its infancy and commercial schemes are not yet in existence, it is not known how their dynamics may be capable of responding to signals derived from frequency deviations.

3.6.4 Photovoltaics

Here a distinction should be made between large concentrated PV installations and numerous roof top systems. For large installations, the comments on wind power apply albeit with some reservations. As the PV systems are interfaced to the grids through power electronic converters and as no mechanical inertia is involved, the speed of response in increments or decrements in power flow can be very fast indeed. On the other hand, solar radiation tends to vary more slowly than wind in the short term, and is fairly predictable. As with other renewables, in the future, PV systems may be required to operate at part load, thus providing 'headroom' for continuous or 'occasional' frequency control.

At this stage of PV technology, roof installations are not yet numerous enough to provide a credible frequency response service. However, in years to come if, as predicted, costs plummet and installations are numbered in millions, it is conceivable that the local inverters are fitted with sophisticated controllers to assist system frequency stability.

3.7 Frequency Control Modelling [21]

3.7.1 Background

Simple lumped parameter modelling is able to provide a rough estimate of the effect on frequency control of feeding relatively large amounts of renewable energy into a power system. A system model of this sort is outlined below, together with results showing the dynamic impact of high penetration of wind energy, and also the way in which dynamic control of demand can be used to good effect.

Modelling a Generator

The dynamics of real generator sets are highly complex and differ considerably between sets. However, it has been shown that a governor with a droop characteristic can be usefully modelled as a proportional controller [22].

The first step is to calculate the generator's target power output, P_{TAR}, using the 4% droop characteristic:

$$P_{TAR} = \left(\frac{f_{SP} - f}{0.04 \times f_{NOM}} \right) P_{MAX} \tag{3.4}$$

where f_{SP} is the generator's set point in Hz, f is the current grid frequency, f_{NOM} is the nominal grid frequency and P_{MAX} is the generator's capacity in MW. The next step is to reduce the error proportionally over time between P_{TAR} and the actual output P at time t, using

$$P_{(t+dT)} = P_{(t)} + (P_{TAR} - P_{(t)}) G dT \tag{3.5}$$

where G is the governor gain. It can be shown that an appropriate value for G is about 0.3 as this results in a realistic settling time in frequency of the order of 15 to 20 seconds after a step-change in load.

To get approximate but useful results it has been shown [22] that the total amount of primary response on the system can be modelled by a single governor-controlled generator of sufficient size to represent all generators with headroom. Also, the total amount of base generation can be modelled by an additional very large generator, but on fixed full output.

Modelling Released Demand

Many loads on the grid consist of rotating machines. As mentioned in Section 3.3.5 there is a built-in frequency dependence caused by the fact that these machines slow down as the frequency drops, and thus consume less power. It has been found empirically that for the UK the total active power demand decreases by 1–2% for a 1% fall in frequency depending on the load damping constant, D [22]. This change in power is the released demand. It is treated in the simulation as an injection of active power, P_R, given by

$$P_R = -DP_L \left(\frac{f - f_{NOM}}{f_{NOM}} \right) \tag{3.6}$$

where D for the UK is assumed here to be 1.0 and P_L is the total load if no built-in frequency dependence exists.

Modelling the Grid's Inertial Energy Store

As already stated, the grid frequency falls as all the spinning machines on the system begin to slow down. In effect, the demand deficit is being met by extracting energy from the rotational inertia of all the generators (and spinning loads). The fall in frequency will continue until the demand deficit is met by a combination of released demand and increased generation due to the governor response.

In the simulation, all the inertia is assumed to be stored in an equivalent single flywheel of moment of inertia, I, rotating at grid frequency, $\omega = 2\pi f \mathrm{rad/s}$. The total kinetic energy stored is therefore

$$\mathrm{KE} = \frac{1}{2} I \omega^2 \qquad (3.7)$$

The inertial storage capacity of a power system is measured by an inertia constant, H, which is the number of full-output seconds of energy stored (assuming nominal frequency).

Therefore,

$$H P_{\mathrm{GMAX}} = \mathrm{KE} = \frac{1}{2} I \omega^2$$

where P_{GMAX} is the total generation capacity. H is typically within the range of 2–8 seconds [22]. For this study, H is assumed to be 4. The inertia I for the system is calculated once at the start of the run:

$$I = \frac{2 P_{\mathrm{GMAX}} H}{\omega_{\mathrm{NOM}}^2} \qquad (3.8)$$

For each step of the simulation, the total power surplus, P_{S} is then calculated:

$$P_{\mathrm{S}} = P_{\mathrm{G}} + P_{\mathrm{R}} - P_{\mathrm{L}} \qquad (3.9)$$

where P_{G} is the total generation, P_{R} is the released demand and P_{L} is the load. Clearly, P_{S} is the power going into the inertial energy store. Given that for each simulation time slice, dT, energy must be conserved, then

$$\mathrm{KE}_{(t+dT)} = \mathrm{KE}_{(t)} + P_{\mathrm{S}} dT \qquad (3.10)$$

Hence

$$\frac{1}{2} I \omega_{(t+dT)}^2 = \frac{1}{2} I \omega_{(t)}^2 + P_{\mathrm{S}} dT \qquad (3.11)$$

which provides a difference equation for calculating the new frequency for each step of the simulation:

$$\omega_{(t+dT)} = \sqrt{\omega_{(t)}^2 + \frac{2 P_{\mathrm{S}} dT}{I}} \qquad (3.12)$$

The simulation is then carried through the following steps for each time slice, dT:

1. Calculate P_L by summing the connected loads.
2. Calculate P_G by summing the total generation.
3. Calculate P_R using Equation (3.6).
4. Calculate P_S using Equation (3.9).
5. Calculate the new ω (and f) using Equation (3.12).

3.7.2 A Modelling Example

The model described above was used to assess the effect that a large wind power input would have on the frequency stability of a power system. The example simulated is extreme and has been chosen because it illustrates key issues.

Wind speed data from 23 UK sites were used in the simulation. A 50 hour period containing exceptional wind variability was chosen so as to provide a major challenge to integration. It was assumed that the variation in wind speed and physical separation of the wind turbines in each site would smooth second to second variations in power and thus the power system could be adequately modelled on a minute to minute basis.

For each site, the power output was calculated on the assumption that a wind farm comprising 150 4 MW variable speed wind turbines was present at each site. A purely cubic power–wind speed relationship was assumed with a cut-in wind speed of 2 m/s, a rated wind speed of 15 m/s and a cut-out speed of 25 m/s. The output power from the 23 sites were added together to give a total maximum generation capacity of 13.8 GW. This represents a level of penetration of 25% as a fraction of peak demand.

The simulation results are shown in Figure 3.20. The maximum power reached in the 50 hours chosen was 4.6 GW and the minimum was 550 MW. During the period, the largest sustained drop in wind power occurred during the 37th hour when 5.5 GW of wind was lost in 4.5 hours.

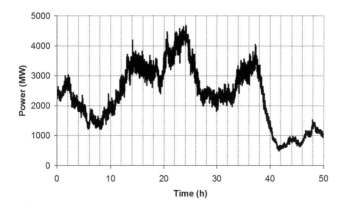

Figure 3.20   Simulated wind power using measured wind speed data from 23 UK sites

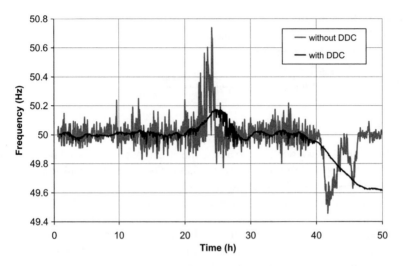

Figure 3.21   Simulation of frequency with wind on the system

A simulation was set up with the above wind generation connected, along with 3170 MW reserve (providing both primary and secondary response), and a frequency dependent load of 36 000 MW at 50 Hz. The level of reserve was chosen arbitrarily so that the maximum frequency excursion would remain approximately within the statutory limits. Enough base generation was provided such that the reserve was half-loaded when the wind power was at its average for the 50 hour period – approximately 36 000 − 2500 − (3170/2) ≈ 32 000 MW.

Wind predictions were not used and the reserve requirements are large enough to compensate for substantial but untypical large positive or negative variations of the wind resource over several hours. The resulting grid frequency–time relationship for this simulation is shown by the curve 'without DDC (dynamic demand control, see Section 3.8.3)' in Figure 3.21.

Between 20 and 25 hours, the wind speed increased substantially to the extent that at 24 hours the base load plus the input from the wind exceeded the nominal demand. The consequence was that the frequency increased excessively to a level determined by the frequency dependence of the load. This undesirable situation could have been easily avoided by fitting wind turbines with a frequency dependent power regulator that reduces power generated at higher frequencies.

At the opposite extreme between 40 and 42 hours, the output from the wind farms was so low that the base load plus the total available reserve (the remaining half of 3170 MW) was insufficient to balance the demand and the frequency had to drop to a level determined again by the frequency dependence of the load. This undesirable situation would have been prevented if at 37 hours use was made of meteorological data predicting a power decline from the wind. Standby generating plant would then have been commissioned to provide the expected shortfall in generation.

3.8 Energy Storage

3.8.1 Introduction

Energy storage devices capable of being topped up during periods of low demand and drained during periods of high demand would aid enormously the frequency control problem. Pumped storage schemes are classical but very expensive examples of such devices. Storage devices, if suitably cheap and efficient could be of benefit to the integration of high levels of RE source penetration, although if such devices were to be present in a power system they would be of operational benefit to traditional energy sources as well.

Income may be derived from an energy store by charging it when the local electricity value is low and discharging it when the value is high, but the level of income critically depends on the round trip efficiency and how this compares with the system electricity cost differential. If, at times, the grid at the point of connection of the embedded renewable generation cannot absorb the entire output of the generator a local storage device could prevent curtailment. Whether this is an economically viable strategy depends, among other things, on the capital cost, the round trip efficiency and the operation and maintenance costs of the device. Another source of income from such devices results from the supply of ancillary services, for example reactive power, voltage and frequency control, and emergency power during a power outage. It is clear that assessing the economic benefit of storage is not a simple matter and that if renewable energy sources have low capital costs it may be more effective to curtail output whenever a surplus exists, rather than attempt to store the surplus.

3.8.2 Storage Devices

As mentioned above, storage technologies depending on their characteristics may aid the integration of renewables, and they could assist the operation/control of a power system over the range of timescales discussed in Section 3.4.6. Conventional technologies include:

1. *The large hydro* is an 'old' renewable but whenever available could facilitate the take-up of variable renewable sources by suitable timing of water release.
2. *Compressed air storage* has been talked about for some time and involves the storage of compressed air in disused underground cavities, e.g. exhausted salt mines. At present it is uncertain how economically and technologically viable this technology is.
3. *Pumped hydro* is an excellent energy storage technique as far as the maturity of technology is concerned, but few attractive sites exist and upfront capital costs are very high.

Less conventional technologies include:

4. *Hydrogen* can be produced by the electrolysis of water using energy from a renewable resource. It can then be 'burnt' [2] as fuel to generate electricity. Alternatively it can be piped as a gas or liquid to consumers to be used locally providing both electricity and heating in a total energy scheme, or it can be used for transport. The combustion of

---

[2] Literally burnt or oxidized in a fuel cell.

hydrogen results in energy plus pure water with no harmful emissions or by-products. This may not be attractive if electricity is the final product since the round trip efficiency is very low (generally below 50%). For this reason there is much interest in using hydrogen for transport, but this depends on much improved on board storage systems.

5. *Flow cells* operate in a mode similar to that of a car battery but without involving the electrodes. Instead, when the flow cell is used as a 'sink', electrical energy is converted into chemical energy by 'charging' two liquid electrolyte solutions. The stored energy can be released on discharge. In common with all DC systems connected to the AC network, a bidirectional power electronic converter is required.

6. *Batteries.* The lead–acid battery is one of the most developed battery technologies. It is a low cost and popular choice of electrical energy storage but has disadvantages in terms of energy density per unit weight, a short cycle life and the dependence of the delivered energy on the rate of discharge. As a consequence, a large variety of other batteries have been under intensive development to provide high round trip efficiency, low life-cycle cost, high reliability, capability of deep discharge and a large number of charge–discharge cycles, low maintenance, high power energy density per unit weight and low capital cost. Although they are generally perceived as too expensive for general inclusion in large power systems, high temperature batteries such as the Zebra cell [23], are of increasing commercial interest at the substation level.

7. *Flywheels.* There has been a number of projects carried out to store energy in very fast rotating flywheels. The major problems that had to be overcome were the maintenance issues and losses relating to bearings, the low specific strength of standard materials such as steel and the associated severe failure management problems at high speeds. Contemporary flywheels are made of fibre-reinforced composites which have powdered magnetic material introduced into the composite which when magnetized form either the rotor of a high speed motor/generator or the rotating element of passive magnetic bearings. The flywheel motor/generator is interfaced to the mains through a power electronic converter. Currently the technology is expensive and only used for niche applications.

8. *Super capacitors*, alternatively known as ultracapacitors, consist of a pair of metal foil electrodes, each of which has an activated carbon material deposited on one side. The activated carbon sides are separated by a paper membrane and then rolled into a package. Ultracapacitor operation relies on an electrostatic effect whereby charging and discharging takes place with the purely physical (not chemical) and reversible movement of ions. As a result there are some fundamental property differences between ultracapacitors and battery technologies including long shelf and operating life as well as large charge–discharge cycles of up to 500 000.

9. *Superconducting magnetic energy storage* (SMES) stores energy within a magnetic field created by the flow of direct current in a coil of superconducting material. Typically, the coil is maintained in its superconducting state through immersion in liquid helium at 4.2 K within a vacuum-insulated cryostat. A power electronic converter interfaces the SMES to the grid and controls the energy flow bidirectionally. With the recent development of materials that exhibit superconductivity closer to room temperatures this technology may become economically viable.

10. *Heat or cold store.* There is a long tradition of using thermal storage to assist in power system operation. The UK's old white metre off-peak tariff and the more recent Economy 7 tariff were both primarily for charging storage heaters. More recently, sophisticated

ways to use heat and cold storage in the context of power systems has been explored. This technology involves modulation of the energy absorbed by individual consumer electric heating elements and refrigeration systems for the benefit of overall system power balance. An example of this is described in the next section.

3.8.3 Dynamic Demand Control

A scheme is being investigated that uses the already existing stored energy in millions of consumer appliances and requires the installation of *dynamic demand control* (DDC). These monitor system frequency and switch the appliance on or off, striking a compromise between the needs of the appliance and the grid. Initially fridges and freezer applications have been investigated.

Refrigerators are 'on' in all seasons, throughout the day and night, and are therefore available to participate in frequency control at all times. The total energy demand on the UK grid from domestic (excluding industrial and commercial) refrigeration has been estimated as 16.7 TW h per year which amounts to an average load of 1.9 GW. The refrigeration load is dependent on ambient temperature, winter load being approximately two-thirds that in summer. Daytime load is also slightly higher than that at night. Refrigerators are designed to handle considerable switching as they typically have a switching cycle of the order of 5 minutes to 1 hour depending on characteristics and contents. Any additional switching caused by frequency control should not therefore present a problem.

As an example, the power system model of Section 3.7.2 was used with the addition of DDC refrigerators. For this purpose, a refrigerator with a dynamic demand controller was modelled. The aggregation of 24.9 million such appliances (one per UK household) with statistically uncorrelated behaviour and equivalent to 1320 MW of deferrable load was investigated for its response to fluctuating wind power with a reserve of 2000 MW. Figure 3.21 shows that the system with the DDC (black trace) considerably reduced the variation in frequency even though the system was operating with substantially less reserve. This is because the simulated controllers reacted more quickly than the generator governor to changes in frequency.

The graph also shows that the system frequency with and without DDC fell below the operational limit of 49.8 Hz. However, the system with DDC provided a considerable breathing time. Frequency did not fall below the operational limit until nearly 2 hours after the non-DDC system. A team of engineers operating the power network would therefore be given a wider choice of generation with which to balance the system including slower acting (and therefore possibly cheaper and more efficient) options. Also the delay may allow generation to be scheduled more cost effectively through the electricity market, which may operate a 'gate closure' (see Chapter 7) time of half an hour in advance of any particular generation slot.

It may be concluded that an aggregation of a large number of dynamically controlled loads has the potential of providing added frequency stability and smoothing to power networks, both at times of sudden increase in demand (or loss of generation) and during times of fluctuating wind or other renewable power. The devices, if incorporated on a real system, could displace some reserve and result in a significant reduction in governor activity of the remaining generators with assigned headroom. The amount of reserve displaced will depend on the extent to which low magnitude but long term frequency excursions can be tolerated, and on the amount of slower-acting back-up generation available, but it could be of the order of the total amount of dynamically controlled loads connected.

The potential demand that could be operated under dynamic control is considerable. Deep-freeze units, industrial and commercial refrigeration, air conditioning as well as water heating systems could provide DDC. In principle, the potential available in a developed country would be several GW. The future will tell whether this potentially useful, but not essential, companion to variable renewable energy sources will attract the attention it deserves.

References

[1] Laithwaite, E.R. and Freris, L.L. *Electric Energy: Its Generation, Transmission and Use*, McGraw-Hill, Maid-enhead, 1980.
[2] 'Ancillary service provision from distributed generation', DTI report, Contract No. DG/CG/00030/00/00, URN No. 04/1738, UK.
[3] 'National Grid Seven Year Statement', available from www.nationalgrid.com/uk/library/documents.
[4] 'The costs and impacts of intermittency', Report by UKERC, Imperial College, London, March 2006.
[5] Farmer, E.D., Newman, V.G. and Ashmole, P.H. 'Economic and operational implications of a complex of wind-driven power generators on a power system', *IEE Proceedings A*, 1980, **127**(5).
[6] Sinden, G. '*Wind power and the UK wind resource*', Environmental Change Institute, University of Oxford, 2005.
[7] Van Hulle, F. 'Large scale integration of wind energy in the European power supply: analysis, issues and rec-ommendations', Report by EWEA, December 2005.
[8] 'ILEX Energy Consulting and UMIST at quantifying the system costs of additional renewables in 2020', A Report of DTI and UMIST, October 2002.
[9] Grubb, M.J. 'Value of variable sources on power systems', *IEE Proceedings C*, March 1991, **138**(2).
[10] Grubb, M.J. 'The integration of renewable energy sources', *Energy Policy*, September 1991.
[11] Parsons, B. and Milligan, M. 'Grid impacts of wind power: a summary of recent studies in the United States', in European Wind Energy Conference, Madrid, June 2003.
[12] Milborrow, D. 'Forecasting for scheduled delivery', *Wind Power Monthly*, December 2003, **19**(12), p.37.
[13] Milborrow, D. 'False alarm', *IEE Power Engineering*, April/May 2005.
[14] Sinden, G. 'The practicalities of developing renewable energy stand-by capacity and intermittency', Submission to the Science and Technology Select Committee of the House of Lords, Environmental Change Institute, University of Oxford.
[15] Kariniotakis, G., Marti, I., *et al.* 'What performance can be expected by short-term wind power prediction models depending on site characteristics?' in Eropean Wind Energy Conference (EWEC 2004), London, 2004.
[16] 'Wind power in the UK', Sustainable Development Commission, UK, 2006.
[17] Sinden, G. 'The practicalities of developing renewable energy. Stand-by capacity and intermittency', Submis-sion to the Science and Technology Committee of the House of Lords' Environmental Change Institute, Uni-versity of Oxford.
[18] Streater, C.J.M. 'Scenarios for supply of 100% of UK energy requirements from renewable sources', REST MSc Thesis, Department of Electronic and Electrical Engineering, Loughborough University, UK.
[19] Barret, 'Integrated systems modelling', Chapter in *Renewable Electricity and the Grid* (ed. G. Boyle), Earths-can, 2007.
[20] 'Planning of the grid integration of wind energy in Germany onshore and off-shore up to the year 2020', in International Conference on *The Integration of Wind Power into the German Electricity Supply*, Berlin, May 2005.
[21] Short, J.A., Infield, D.G., Freris L.L. 'Stabilization of Grid Frequency through Dynamic Demand Control' *IEEE Transactions on Power Systems*, **22**, 3, August 2007.
[22] Kundur, P. *Power System Stability and Control*, McGraw-Hill, Maidenhead.

[23] Tilley, A.R. and Bull, R.N. 'The Zebra electric vehicle battery – recent advances', in Proceedings of the Auto-tech '97 Conference, NEC, Birmingham, UK, 4–6 November 1997.

Other Useful Reading

Barton, J.P. and Infield, D.G. 'Energy Storage and its Use with Intermittent Renewable Energy' *IEEE Transactions on Energy Conversion*, **19**(2), June 2004.

Hartnell, G. 'Wind on the system-grid integration of wind power', *Renewable Energy World*, March–April 2000, **3**(2).

Holt, J.S., Milborrow, D.J. and Thorpe, A. 'Assessment of the impact of wind energy on the CEGB system', CEC Brussels, Contract No. EN3W-0058-UK.

Patterson, W. *Transforming Electricity*, Earthscan, 1999. This book gives a readable, informative and entertaining account of the development of power systems over the past century.

'Wind power in the UK', Report by the Sustainable Development Commission, UK, May 2005.

4

Electrical Power Generation and Conditioning

4.1 The Conversion of Renewable Energy into Electrical Form

Renewable energy is available in a variety of forms. Uniquely, biomass is available in the form of combustible fuel and thus can play a similar role in generation as conventional fossil fuels. To generate electricity, all other renewables require a number of conversion stages that differ from those found in systems based on traditional fuels.

- Wind energy is available in kinetic form. The function of a wind turbine is to extract energy from the intercepted wind by slowing it down and to convert this energy into a mechanical form that suits an electric generator. To improve the efficiency of conversion and for other operational reasons, the generator may be interfaced to the mains through a power electronic converter.
- Water power in the form of tides (potential) or water flow (kinetic) requires a turbine to transform this energy into rotational form for further conversion into electricity, again through a generator.
- The kinetic energy in water currents caused by tidal effects can alternatively be captured through an underwater 'wind turbine' that uses the same technology as those on land.
- Wave energy conversion requires specially designed devices that transform the low frequency energy in the waves into (usually) pressure energy in oil, which in turn drives an electrical generator. In other wave energy concepts the rise and fall of waves drives air through a turbine coupled to an electrical generator.
- Solar energy is available as radiation ranging from ultraviolet to infrared. Conversion into electricity can be implemented thermally by solar furnaces that raise steam to drive conventional turbines or through solid state photovoltaic devices that utilize the radiation to separate charges in semiconductor junctions. The efficient operation of PV based systems depends on interfacing the PV array to the grid through a power electronic converter.

Renewable Energy in Power Systems   Leon Freris and David Infield
© 2008 John Wiley & Sons, Ltd

This summary indicates that all renewables, except PV systems, rely on electromechanical generators for the final stage of conversion from mechanical into electrical energy. This chapter introduces the principles of operation of two classes of electrical generators, the 'synchronous' and the 'asynchronous' types, both used extensively in RE applications. Additionally, this chapter deals briefly with the principle of operation of the transformer, a ubiquitous device in multivoltage level power systems. Understanding the operation of the transformer is a necessary prerequisite for the study of the 'asynchronous' type of generator.

Power electronics plays a vital role in PV and an increasingly important role in the wind power area. A review of power electronic devices and the converters based on them is covered in the penultimate section. Finally, the chapter concludes with a description of how electro-mechanical and/or power electronic converters are used in PV and wind systems.

In what follows a symbol written in regular type indicates that the parameter is a scalar while bold type is used if it is a vector, phasor or a complex number.

4.2 The Synchronous Generator

4.2.1 Construction and Mode of Operation

In an electrical generator, mechanical input power is converted into electrical output power. To get an appreciation of how this energy conversion process is carried out it is useful to look briefly at the underpinning physics. Faraday's law of electromagnetic induction [1] states that a conductor of length l (m) moving with a velocity u (m/s) through a magnetic field of uniform flux density B (Tesla), l, u and B being mutually perpendicular, will experience an induced voltage across it given by

$$v = Blu \quad \text{(volt)} \tag{4.1}$$

Nature is such that this mechanical (u) to electrical (v) conversion process described by Equation (4.1) is 'mediated' through the presence of a magnetic field (B). The equation shows that to generate a high and therefore useful voltage it is necessary to have a high magnetic flux density, a long conductor length and as high a conductor velocity as possible. All the above requirements are particularly well satisfied if the mechanical–magnetic–electrical structure of the generator is arranged as a rotating rather than a linear one. In practice it has been found that it is preferable to have the conductors stationary and move the source of the magnetic field. With a stationary set of conductors the problems of insulation and electrical connections are eased and centrifugal forces on the main windings are absent.

Figure 4.1(a) shows in outline an AC generator also known as an *alternator* or *synchronous generator*. Here the conductors that form a winding, known as the *stator*, are stationary and the source of magnetic flux rotates. The source of flux is a rotor with poles marked north and south that carries a *field winding* as shown in the figure. The field winding is fed or *excited* from an external DC source through sliding contacts known as *slip-rings*. An additional reason why it is preferable to arrange the main windings to be stationary is that the DC power associated with the field is a small fraction of the power delivered by the stator winding. In fact, in some designs the field winding can be dispensed with completely and replaced by a

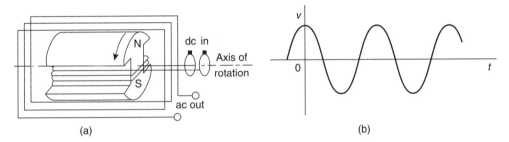

Figure 4.1   A two-pole synchronous generator. (Reproduced from Reference [1] with permission of John Wiley & Sons, Inc.)

rotor built with permanent magnets. There are both advantages and disadvantages to each type of excitation which will be explored later.

The flux density produced by the rotor in Figure 4.1(a) is maximum positive upwards along the pole axis, zero at $90°$ to the pole axis and maximum negative at $180°$. Equation (4.1) indicates that a variable voltage with polarity reversals is generated when the field winding rotates at constant angular velocity because the magnetic field cutting each conductor increases to a maximum, decreases and successively reverses in each revolution. By proper spatial distribution of the stator winding turns and shaping of the pole faces, the generated voltage across the stator terminals in Figure 4.1(a) can be made to approach the sinusoid waveform shown in Figure 4.1(b). It should be evident that a full rotation of the rotor will result in a full complete cycle of the sinusoidal waveform. Hence the frequency of the generated voltage in Hz (cycles per second) is identical to the angular velocity of the rotor in revolutions per second.

Appendix A explains that three-phase AC systems are universally used for the generation, transmission and utilization of electrical energy. One of the reasons is that synchronous generators are particularly well suited for the generation of three-phase voltages. When wound for three phases, alternators make optimal use of the iron that carries the magnetic flux and of copper that carries the electric current. Figure 4.2(a) shows three separate one turn windings, where a, b and c indicate the beginnings and a', b' and c' the ends of these windings. The winding axes are shifted in space with respect to each other by $120°$. As a consequence the voltages generated by the rotating field are also shifted in the time domain by one third of a period, thus forming the three-phase system of AC supply shown in Figure 4.2(b).

A more effective arrangement for superior power output from a three-phase stator winding is shown in Figure 4.3. Here the winding is embedded and distributed in slots in an iron cylinder. The magnetic flux ϕ and therefore the flux density B generated by the rotating electromagnet is approximately proportional to the magnetomotive force (mmf) F (ampere turns) given by

$$F = N \ I_f \tag{4.2}$$

where N is the number of turns of the field winding (shown as having one turn in the figure) and I_f is the field or excitation current. In a practical machine the air gap between the rotor pole faces and the internal surface of the cylinder is made small, and the cylinder itself is made of a ferromagnetic material so that minimum resistance is offered to the flow of magnetic flux. A large radial magnetic flux can therefore be produced from a moderate I_f.

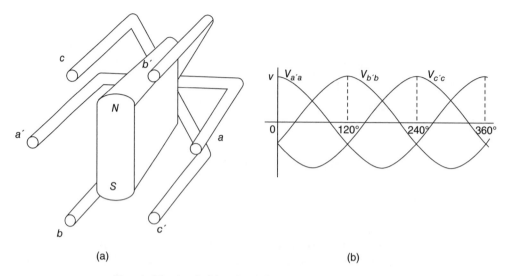

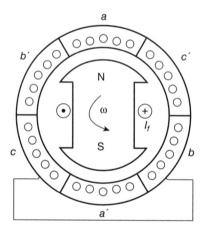

(a)   (b)

Figure 4.2   A primitive three-phase synchronous generator

Figure 4.3   A practical three-phase synchronous generator

The peripheral velocity with which the conductors in the embedded windings cut the magnetic flux is, of course, proportional to the angular velocity ω. Using Equations (4.1) and (4.2), the induced voltage V generated in a stator winding for a rotating generator is therefore given approximately by:

$$V = k\omega I_f \tag{4.3}$$

where k is a constant of proportionality that depends on the number of turns in each winding, the distribution of the conductors in the slots, the length of the air gap and the general geometry and magnetic properties of the iron that carries the magnetic flux. Electrical machine designers take all of these factors into account when designing synchronous generators so that the mechanical input power is converted into electrical power with maximum efficiency and with minimum cost of materials.

Up to this point the alternator in standalone mode has been considered. The rotor driven by a prime mover induces a set of balanced three-phase voltages in the stator windings, as shown in Figure 4.2(b). If a balanced three-phase load is connected across the windings, a balanced set of currents will transfer power from the prime mover to the load. This standalone operation is the exception rather than the rule in power generation. Rather, the alternator is most likely to be required to inject power into a grid that to all intents and purposes may be considered an *infinite bus*. This mode of operation is discussed in the following sections.

4.2.2 The Rotating Magnetic Field

An examination will first be made of what happens when the primitive stator windings of Figure 4.2(a) are connected to a three-phase supply and the rotor is absent. Figure 4.4(a) shows the balanced currents that will be drawn by the three primitive windings which are redrawn in Figure 4.4(b) [1]. When stator winding aa' carries the sinusoidal current i_a, Equation (4.2) tells us that i_a will generate a sinusoidally pulsating space vector field F_a and therefore a flux ϕ_a along the axis of coil aa' in Figure 4.4(b).

Similarly, the remaining two currents i_b and i_c in Figure 4.4(a) will generate pulsating fields and fluxes F_b, ϕ_b and F_c, ϕ_c along the axes of coils bb' and cc' respectively. Positive currents in the windings flowing into the $\otimes$ conductor and returning through the $\odot$ conductor in each

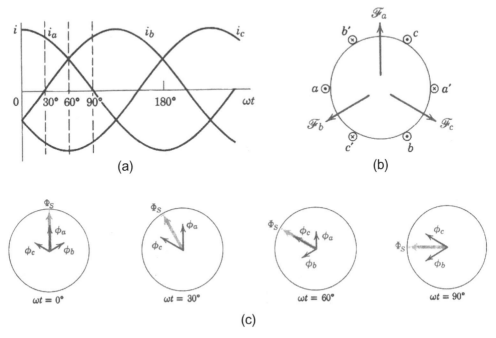

(a)   (b)

(c)

Figure 4.4 Production of a rotating magnetic field. (Reproduced from Reference [1] with permission of John Wiley & Sons, Inc.)

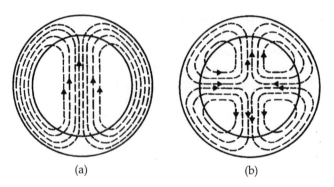

(a)   (b)

Figure 4.5   A one- and two-pole pair wound stators

phase produce mmfs F_a, F_b and F_c in the directions indicated in Figure 4.4(b). At the instant $\omega t = 0°$ in Figure 4.4(a), i_a is a positive maximum and i_b and i_c are negative and one-half maximum. At this instant, the flux contributions in space can now be added as shown in Figure 4.4(c) for $\omega t = 0°$. The stator flux Φ_s is the space vector sum of the three flux contributions. At $\omega t = 30°$, the relative magnitudes of the three currents have changed and the position of the resulting stator flux has shifted anticlockwise by $30°$. At $\omega t = 60°$, the current magnitudes have changed again and the resultant stator flux vector has shifted another $30°$. Over a complete cycle of mains frequency the vector Φ_s would have completed one revolution. It follows that for a 50 Hz supply the flux vector will complete one revolution in one-fiftieth of a second; i.e. it will complete 50 revolutions per second. It is said that a rotating magnetic field (RMF) has been created by virtue of the space distribution of the three windings and of the time distribution of the currents in the three windings. This is a very important concept on which the operation of three-phase generators (and motors) depends.

To summarize, if a stationary observer were to position him- or herself inside the stator cylindrical space of Figure 4.3, he or she will observe a north–south pole pair rotating at 50 × 60 = 3000 revolutions per minute for a 50 Hz supply and 60 × 60 = 3600 revolutions per minute for a 60 Hz mains. These are known as the synchronous speeds for 50 and 60 Hz systems for a one-*pole pair* winding. Stator windings can be arranged so that not just one pair but several pairs of poles can be created in the interior of the stator cavity. For example, a two-pole pair arrangement will have two separate windings for each phase, each winding side occupying 30 rather than the $60°$ shown in Figure 4.3. It can be shown that with such an arrangement in one-fiftieth of a second the rotating magnetic field in a 50 Hz system will advance by 180 rather than $360°$. The magnetic field produced by an alternator wound for one- and two-pole pairs is shown in Figure 4.5 (a) and (b) respectively.

This property is used by designers of electrical machines to generate rotating magnetic fields that rotate at submultiples of the figures shown above. The reason why such machines are useful will be explained in the following section. The general relationship linking the synchronous speed in radians per second ω_s to frequency f and number of pole pairs p is:

$$\omega_s = \frac{2\pi f}{p} \tag{4.4}$$

Table 4.1 Relationship between pole pairs and synchronous speed in rev/min

p	Rev/min (50 Hz)	Rev/min (60 Hz)
1	3000	3600
2	1500	1800
3	1000	1200
4	750	900
⋮	⋮	⋮
200	15	18

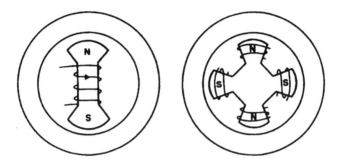

Figure 4.6 Rotor configurations for a one- and two-pole pair wound stators

Table 4.1 shows the relationship between the synchronous speed in rev/min and the number of pole pairs p for 50 and 60 Hz systems.

4.2.3 Synchronous Generator Operation when Grid-Connected

Having established the presence of a rotating magnetic field within the cylindrical interior of the stator, the rotor is now ready to be incorporated. Figure 4.6 shows the salient rotor configurations for a one- and two-pole pair wound stators corresponding to the flux patterns in Figure 4.5.

A one-pole pair rotor will be considered. The rotor field current is switched on so that a north–south pole pair is created. With the rotor stationary, the interaction between the rotating magnetic field (RMF) and the rotor field is to say the least unproductive. As the RMF sweeps by at 50 rev/s or 20 ms per revolution, the only effect experienced by the rotor body is a pulsating torque as the RMF pole pair approaches and then overtakes the rotor pole pair. Due to the rotor inertia it is just not possible for the rotor to accelerate and lock on to the RMF within the required milliseconds.

Consider now that some external torque is applied to drive the rotor at increasing speed until the synchronous speed is very nearly reached. If the speed difference is close enough and decreasing, at some point the rotor N–S pair will lock on to the RMF S–N pair and the two magnet systems will rotate in synchronism. With the external torque now removed and assuming a lossless environment with no friction and windage, the rotor will rotate at synchronous speed with no requirement that any power is fed to it either mechanically from the shaft or electrically from the power network to which the synchronous machine is connected. The system is *idling* and can be thought as being suspended between motoring and generating.

In the synchronous machine the speed of the locked magnet pair is fixed because the infinite bus frequency and therefore the RMF speed is fixed. However, the relative angular disposition of the magnetic axes of the two locked pairs is not fixed and is the mechanism that regulates the direction of energy conversion. If an external braking torque Q_t is applied to the shaft the rotor will keep rotating at the synchronous angular speed ω_s, but its magnetic axis will fall back or spatially lag the magnetic axis of the RMF. The tangential magnetic tension forces caused by the misalignment of the magnet pair is at heart the mechanism of electromechanical energy conversion. The mechanical power P_m given by

$$P_m = Q_t \omega_s \tag{4.5}$$

is extracted from the shaft; therefore an equal amount of electrical power must be supplied from the electrical system to which the machine is connected if the conservation of energy principle is to be satisfied. The synchronous machine is now motoring.

If the external torque is accelerating rather than decelerating the magnetic axis of the rotor is advanced with respect to the RMF axis, mechanical power is now supplied to the shaft and the energy conservation principle demands that an equal amount of power is fed into the electrical grid system. The synchronous machine is now generating.

In renewable energy applications, a high number of pole pairs may be selected for the generator as this requires a low rotational shaft speed to generate mains frequency voltages. Hydro turbines and wind turbines are in this class. The multipole arrangement is particularly desirable when a wind turbine is coupled directly to a synchronous machine without an intervening gearbox. In such cases, if a frequency of 50 Hz is to be generated from a wind turbine rotating at 15 rev/min, Table 4.1 indicates that two hundred pole pairs would be necessary!

This power generation process will now be looked at from the perspective of the electrical network to which the generator is connected.

4.2.4 The Synchronous Generator Equivalent Circuit

To analyse the power flows in electrical systems, component representations are required that can be incorporated into network or circuit models. To use the available circuit analysis tools described in Chapter 5, it is necessary to build up these representations from basic circuit elements, namely: resistors, inductors, capacitors and voltage or current sources (Appendix A). Electrical power engineers over the years have developed a range of what are known as *equivalent circuits* for network simulation of electrical generators.

Here, for an approximate steady state analysis of power flows a description is required of a synchronous generator by the simplest possible equivalent circuit. The Thévenin principle explained in Appendix A can be used, for example, to describe the behaviour of a DC battery by a source voltage in series with a resistance. Amazingly, this principle can also be effectively used to describe through a simple circuit, and to a good approximation, the behaviour

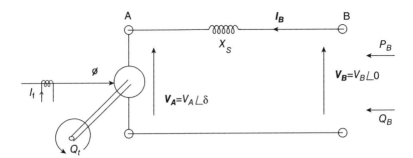

Figure 4.7   Equivalent circuit of a synchronous machine

of a generator as complex as an alternator. The steps in this transformation are not given here but can be found in any book on electrical machines (for example see Reference [2]). In what follows it is assumed that the reader is familiar with the use of phasors to represent AC quantities. Readers not familiar with this concept should, at this stage, study the material in the Appendix.

In Figure 4.7 the electrical generator has been reduced into a single-phase (the relationship with 3-phase is dealt with later) Thevenin equivalent circuit consisting of a voltage source $V_A = V_A \angle \delta$ (the generated or 'internal' voltage of Equation (4.3)) and a source impedance X_s, known as the *synchronous reactance*. The synchronous reactance represents in one lumped element all the internal complex interactions between the rotor and stator magnetic fields, which are not of concern here. To maximize conversion efficiency, synchronous machines are designed to have as low winding resistance as possible; hence the source resistance representing the ohmic value of the stator winding is omitted here with little loss in accuracy. The equivalent circuit is shown connected to an infinite bus, i.e. a network of fixed frequency f and of fixed voltage $V_B = V_B \angle 0°$ where its $0°$ angle defines it as the reference voltage.

An investigation will explore how the two available external control parameters, namely the field current I_f and the shaft torque Q_t, influence the synchronous machine and consequently the equivalent circuit behaviour. Equation (4.3) shows that $|V_A|$ depends on the field current, which is the source of the magnetic flux. It is also known that the angular disposition of the rotor magnetic axis depends on the direction and magnitude of the torque applied to the shaft. Angle δ (the load angle) is defined as the angle by which the axis of the rotor flux space vector that induces V_A leads the axis of the net flux space vector in the machine that induces V_B. The load angle in the spatial disposition of rotating vector fields is the same as that in the phasorial disposition of voltages in the equivalent circuit. An accelerating or 'generating' torque will result in a positive δ and in V_A leading V_B. A decelerating or 'motoring' torque will result in a negative δ and in V_A lagging V_B.

4.2.5 Power Transfer Equations

There is interest in exploring the mechanism by which power is injected into the grid by a synchronous generator. This can be done by means of the concept of complex power developed in Appendix A. The grid connected synchronous generator of Figure 4.7 will be considered. The complex power at end B of the line is given by

$$S_B = V_B I_B^*$$ (4.6)

Then I_B is expressed as a function of the line voltages using Kirchoff's voltage law,

$$I_B = \frac{V_B - V_A}{X_S} \tag{4.7}$$

Substituting Equation (4.7) into Equation (4.6) gives

$$S_B = V_B I_B^* = V_B \left(\frac{V_B - V_A}{X_s} \right)^* = V_B \left(\frac{V_B - V_A e^{-j\delta}}{-jX_s} \right) = j\frac{V_B^2}{X_s} - j\frac{V_A V_B}{X_s} e^{-j\delta} =$$

$$S_B = -\frac{V_A V_B}{X_s} \sin\delta + j\left(\frac{V_B}{X_s}(V_B - V_A \cos\delta) \right) = P_B + jQ_B \tag{4.7a}$$

Hence

$$P_B = -\frac{V_A V_B}{X_s} \sin\delta \tag{4.8a}$$

$$Q_B = \frac{V_B}{X_s}(V_B - V_A \cos\delta) \tag{4.8b}$$

If the above analysis were to be carried out for terminals A in Figure 4.7, the results would be

$$P_A = -P_B \tag{4.9a}$$

and

$$Q_A = \frac{V_A}{X_s}(V_A - V_B \cos\delta) \tag{4.9b}$$

Equation (4.9a) confirms the trivial fact that as the system is lossless, power P_B *coming out* of terminals B is equal to power P_A *fed into* terminals A. The scalar equations (4.8) and (4.9) are important in power systems technology as they describe the flow of active and reactive power of grid-connected synchronous generators.

4.2.6 Three-phase Equations

The synchronous machine equations were derived without any reference to its three-phase nature. Assuming that the voltages in Equations (4.8) and (4.9) are phase to neutral in volts, the equations will give the single-phase active and reactive powers in watts and VAR respectively. If the voltages are in kV then the active and reactive powers – both functions of voltage squared – will be in MW and MVAR.

In a balanced three-phase system the three-phase P and Q will be three times the per-phase P and Q. Applying this to Equation (4.8),

$$P_{3\varphi} = \frac{3V_A V_B}{X_s} \sin\delta = \frac{\sqrt{3}V_A \sqrt{3}V_B}{X_s} \sin\delta = \frac{V_{Al} V_{Bl}}{X_s} \sin\delta$$

where V_{Al} and V_{Bl} are line voltages.

Worked Example 4.1

A synchronous generator of 10 ohms synchronous reactance is supplying 5 MW and 2 MVAR to an 11 kV network. Calculate the generator internal voltage.

Rearranging Equations (4.8a) and (4.8b) gives

$$V_A \sin \delta = \frac{XP}{V_B}$$

and

$$V_A \cos \delta = V_B - \frac{XQ}{V_B}$$

Therefore

$$\tan \delta = \frac{XP}{V_B^2 - XQ}$$

In this case P = –5, Q = –2, V_B = 11, X = 10.

Hence

$$\delta = 19.52°, \text{ and } V_A = 13.6 \,\text{kV}$$

4.2.7 Four-Quadrant Operation

Consider the case where a lossless synchronous machine is run up to synchronous speed by an external prime mover. The field current is then gradually increased until the terminal voltage of the machine (the same as the internal generated voltage, as no current is taken) is made to be equal to the voltage of the local bus to which the machine is to be connected. Precise adjustments to the speed of the prime mover are made so that through some external instrumentation it is possible to detect an instant at which the internal voltage V_A exactly matches V_B in magnitude and phase. The synchronous machine can now be safely connected to the bus. This process is known as *synchronisation* and must be carried out each time a synchronous machine is to be connected to the mains.

Now arrange for the prime mover to apply an accelerating torque. This will result in a positive load angle δ and according to Equation (4.8a) a negative active power, i.e. active power injected into the power system. As expected, the machine is generating. With a braking torque on the shaft, i.e. with the machine motoring, the load angle is negative and active power is supplied to the machine from the power system.

Returning to the idling state, consider now what happens if the field current is increased so that V_A is made larger than V_B, but no external torque is applied so that δ and the active power are zero. Equation (4.8b) shows that the reactive power is negative; i.e. the synchronous machine injects reactive power into the system, and acts as a generator of reactive power. In this state the machine is said to be *overexcited*.

Conversely, if the excitation current is decreased so that $V_A < V_B$, the reactive power flow is reversed, the machine is a consumer of reactive power and it is said to be *underexcited*.

The synchronous machine is capable of operating in each of the four quadrants of the quadrant diagram shown in Figure A18 in the Appendix.

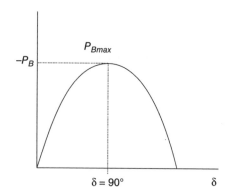

Figure 4.8   Power angle characteristic of an SG

4.2.8 Power – Load Angle Characteristic: Stability

Equation (4.8a) is plotted in Figure 4.8 to illustrate the dependence of the generated active power on the load angle. Note that a synchronous generator (SG) connected to an infinite bus is capable of generating a maximum active power P_{Bmax}, at $\delta = 90°$. Any additional applied mechanical torque will increase the load angle beyond 90°, with a consequential decrease in electrical power. Physically, the peripheral magnetic forces linking the two fields together is insufficient to maintain the locking effect. The power balance between mechanical and electrical powers has now been lost, the excess mechanical power accelerates the rotor beyond synchronous speed and the synchronous generator has lost its ability to act as a stable power converter. In this unstable regime the SG is described as having *lost synchronism* i.e. its rotor generates an internal voltage V_A of a higher frequency than the infinite bus voltage V_B. Operation under this condition, known as *pole slipping*, results in large overcurrents, is highly undesirable and protection equipment will be brought into action to disconnect the synchronous machine from the mains.

During system contingencies violent transient changes may take place that cause swings in the load angles of synchronous generators. To ensure safe stability margins so that pole slipping is prevented, the steady state load angles of SGs are kept well below 30°. Power system stability is a major topic in its own right involving the simultaneous solution of the differential equations characterizing all the network components. This is beyond the scope of this book, but it is worth noting that connection of large RE sources on to the grid will alter the system dynamics. Therefore studies may be required to assess the new system stability margins.

4.3 The Transformer

4.3.1 Transformer Basics

The transformer is an indispensable part of any power system operating at a range of voltages. The transformer mode of operation is included here as it provides a useful aid to the under-

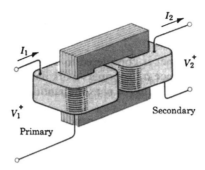

Figure 4.9   The Transformer. (Reproduced from Reference [1] with permission of John Wiley & Sons, Inc.)

standing of the mode of operation of induction generators which are used extensively in wind turbines.

The properties of an inductance are discussed in the Appendix. A current carrying inductor, usually wound in the form of a coil, generates a magnetic flux. This flux links with the turns of the coil or the *winding* and induces a voltage in these turns if the current and therefore the flux is increased or decreased. It stands to reason that if a changing magnetic field produced by one inductor were to 'link' with the turns of wire in another adjacent inductor, it would induce a voltage in the *unpowered* inductor. This phenomenon is called *mutual inductance* and the *transformer* is a device constructed to exploit this effect.

By having two inductors coupled together by a common magnetic field path, it is possible to transfer energy from one inductor circuit to the other. In order for this to work, the magnetic field has to be constantly changing in strength, otherwise no voltage will be induced in the unpowered winding. Thus, the transformer is essentially an AC device. The powered winding of a transformer is called the *primary*, while the unpowered winding is called the *secondary*.

Figure 4.9 shows a transformer in outline [1]. The primary winding is connected to an AC supply V_1. The winding is wound round a substantial closed ferromagnetic core. This provides a very low 'magnetic resistance' to the flow of magnetic flux in comparison to the surrounding air. Because of the very low resistance to the flux flow, only a very low current, known as the *magnetizing current*, is required from the supply to set up an mmf that circulates a flux which induces a voltage in the primary that exactly balances V_1. For the present the effect of this low current may be disregarded. Because of the low 'magnetic resistance' of the ferromagnetic core it can be assumed that practically all of the generated magnetic flux is constrained to flow within the core with only a small amount of leakage flux taking paths in the surrounding air, which will also be disregarded at this point. The consequence of this is that the secondary winding which is also wound round the transformer core, links exactly the same flux as the primary winding. Finally, if it is assumed that the windings consisted of large cross-section copper wire their ohmic resistance could be disregarded. All these simplifications lead to the concept of the *ideal transformer*.

In an ideal transformer the voltage induced in an inductor of N turns linking a flux varying at $d\phi/dt$ is proportional to N. Hence for the transformer in Figure 4.9, we can write

$$\frac{V_1}{V_2} = \frac{N_1}{N_2} \qquad (4.10)$$

Suppose that the secondary is connected to a resistor that draws current I_2 with the consequence that power V_2I_2 is extracted from the secondary. The energy conservation principle requires that this power is supplied from the source to which the primary is connected. It follows that a primary current I_1 is established the value of which can be determined from:

$$V_1I_1 = V_2I_2 \qquad (4.11)$$

and from Equation (4.10)

$$\frac{I_1}{I_2} = \frac{N_2}{N_1} \qquad (4.12)$$

Note that an open circuit or a short circuit on the secondary winding of an ideal transformer are seen as an open or short circuit respectively on the primary side.

An ideal transformer with identical primary and secondary windings would manifest equal voltage and current in both sets of windings. In a perfect world, transformers would transfer electrical power from primary to secondary as efficiently as though the load were directly connected to the primary power source, with no transformer there at all, but it will be found later that this ideal goal cannot be realized in practice. Nevertheless, transformers are highly efficient power transfer devices with no moving parts achieving efficiencies in the high nineties.

The transformer has made long distance transmission of electric power a practical reality, as AC voltage can be 'stepped up' and current 'stepped down' for reduced ohmic resistance losses along power lines connecting generating stations with loads. A transformer that increases voltage from primary to secondary (more secondary winding turns than primary winding turns) is called a *step-up* transformer. Conversely, a transformer designed to do just the opposite is called a *step-down* transformer.

4.3.2 The Transformer Equivalent Circuit

The next task is to develop an equivalent circuit capable of representing realistically the behaviour of a transformer in studies aimed at determining the flow of power in electrical systems. Figure 4.10 shows how such an equivalent circuit can be built up starting with the ideal transformer in the hatched box. The small but finite current required to set up the flux in the core is simulated by the presence of the shunt inductance L_m, the *magnetizing inductance*, which has a large value in henries.

Next, the finite resistance of the two transformer windings can be simulated by the series resistors R_1 and R_2. To keep losses low, these have low numerical values. Finally, any magnetic flux not contained in the core is free to store and release energy rather than transfer it from one coil to the other. Any energy thus stored by this *uncoupled* flux manifests itself as an inductance in series with the relevant winding. This stray inductance is called *leakage inductance* and is represented in the equivalent circuit by L_1 and L_2 both having numerical values much smaller than L_m.

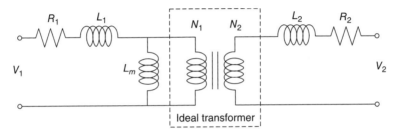

Figure 4.10 The transformer equivalent circuit

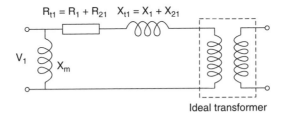

Figure 4.11 Simplified transformer equivalent circuit

The equivalent circuit of Figure 4.10 can be further simplified if the R_2 and I_2 are transferred to the primary so that the primary and secondary resistances and inductances could be lumped into just two series components. This can be done through the conservation of energy principle. The transferred resistance from the secondary (2) to primary (1), which can be called R_{21}, should have such a value that when it carries I_1 should dissipate the same amount of power as R_2 when carrying I_2, i.e. $R_2 I_2^2 = R_{21} I_1^2$. Substituting the number of turns for currents from Equation (4.12) we get:

$$R_{21} = R_2 \left(\frac{N_1}{N_2} \right)^2 \tag{4.13}$$

Similar logic based on the reactive power conservation principle (Appendix A) can be applied for the transfer of the secondary winding reactance to the primary:

$$X_{21} = X_2 \left(\frac{N_1}{N_2} \right)^2 \tag{4.14}$$

A new equivalent circuit incorporating these changes is shown in Figure 4.11. Here R_{t1} and X_{t1} are the total winding resistance and reactance respectively referred to the primary winding. In this circuit the magnetizing reactance X_m is connected across the mains with little loss in accuracy because it can be shown that the voltage drops across R_{t1} and X_{t1} are small. This equivalent circuit is frequently used to calculate the effect of a transformer in a power network and will be used later to develop an equivalent circuit for the asynchronous generator.

4.3.3 Further Details on Transformers

The manufacturer of an electrical machine such as a transformer will indicate on the name-plate the normal operating conditions, e.g.: '11 000 : 415 V, 50 Hz, 500 kVA'. The *rated output* of 500 kVA can be maintained continuously without excessive heating and the consequential deterioration of the winding insulation. Because the heating is dependent on the square of the current, the output is rated in apparent power (kVA) rather than active power (kW). When supplying a zero power factor load, a transformer can be operating at rated temperature while delivering zero active power.

On large transformers, taps on the windings allow small adjustments on the turns ratio. Often these taps are operated by an automatic tap changer that maintains the voltage, usually on the secondary, at a fixed value irrespective of the load on the transformer.

It can be shown [1] that the size and therefore, weight and cost of a transformer are inti-mately related to the frequency of operation. The higher the frequency the lower are the weight and cost. For these reasons, in power electronic systems whenever a transformer is to be used, higher frequencies are employed, a topic to be revisited later.

4.4 The Asynchronous Generator

4.4.1 Construction and Properties

Asynchronous or *induction machines* operating as motors are the most widely used electro-mechanical converters. In an induction machine the stator is identical to the one for synchro-nous machines shown in Figure 4.3 in which three-phase currents supplied to the stator produce a rotating magnetic field (RMF). The rotor, however, is radically different and it has neither an external magnetizing source nor permanent magnets. Instead, alternating currents are injected in the rotor from the stator through induction or transformer action – hence the useful parallel with the operation of a transformer. It is the interaction between these induced rotor currents and the stator RMF that results in torque production.

In its most common form, the rotor consists of axial conductors shorted at the ends by circular rings to form a *squirrel-cage* or just *cage*, as shown in Figure 4.12. Although for the purposes of renewable energy sources there is interest in the generation mode, it is easier initially to understand the operation of the induction machine from the motoring perspective.

As the stator RMF moves at ω_s (given by Equation (4.4)) past the stationary rotor conduc-tors, three-phase electromotive forces (EMFs) are induced in the spatially shifted rotor con-ductors by a flux cutting action. The resulting rotor currents, according to Lenz's law [1], are of such magnitude and direction as to generate a torque that speeds up the rotor. If the rotor were to achieve speed ω_s, there would be no change in flux linkage, no induced voltage, no current in the rotor conductors and therefore no torque. For EMFs to be induced in the rotor conductors they should possess some relative speed with respect to the stator RMF. For motoring, the rotor therefore turns at a lower speed ω_r.

It can be shown that the rotor currents produce an RMF whose speed depends on the fre-quency of these currents. For a constant torque interaction to take place, the rotor RMF must rotate in synchronism with the stator RMF, as in the case of the synchronous machine. How is this accomplished if the rotor rotates at a lower speed than ω_s?

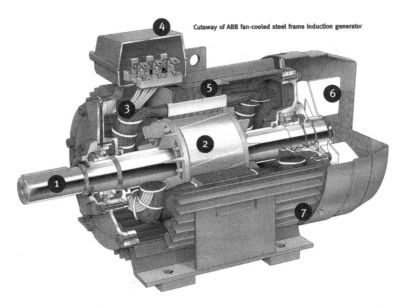

Figure 4.12   Induction machine with cage rotor (1, shaft; 2, cage rotor; 3, stator three-phase winding; 4, terminal box; 5, stator iron core; 6, cooling fan; 7, motor frame). (Reproduced with permission of Asea Brown Boveri Ltd)

The difference between ω_s and ω_r is expressed as a ratio with respect to ω_s and is known as *slip s* where

$$s = \frac{(\omega_s - \omega_r)}{\omega_s} \tag{4.15}$$

Therefore

$$\omega_r = (1-s)\omega_s \tag{4.16}$$

The relative motion between the stator's and rotor's RMFs determines how frequently the stator RMF cuts the rotating rotor conductors, so the frequency of the rotor induced voltages and currents f_r is

$$f_r = s\ f \tag{4.17}$$

where f is the mains frequency. The frequency of the rotor currents determines ω_{rr}, the speed of the rotor RMF with respect to the rotor:

$$\omega_{rr} = 2\pi\ f_r / p = 2\pi\ s\ f / p = s\omega_s \tag{4.18}$$

The speed of the rotor RMF with respect to the stationary stator is the rotor speed plus the rotor RMF's speed with respect to the rotor:

$$\omega_r + \omega_{rr} = (1-s)\ \omega_s + s\ \omega_s = \omega_s$$

It can be concluded that the rotor and stator RMFs rotate together at synchronous speed as required for a uniform torque to be developed as in the synchronous machine. However, in contrast to the synchronous machine, the rotor RMF is produced through *induction* from the stator. The larger the applied braking torque the higher the slip, the larger are the induced EMFs and resulting rotor currents, and the stronger the interaction between the two RMFs to produce an electrical torque equal and opposite to the braking torque. The induction motor therefore exhibits a small decrease in speed with increments in braking torque.

An ideal induction machine could be imagined to operate at zero slip. This is equivalent to the 'idling' state of the fixed speed synchronous machine. The vital difference, however, in the induction machine is that motoring or generating torques will be accompanied by a decrease or increase of speed below or above synchronous respectively. For generating, Equation (4.16) now gives a negative slip. The induction machine will move seamlessly from the motoring into the generation mode as the external torque changes from a decelerating to an accelerating type. Indeed, in small wind turbines, it is very common to find that the induction generator was originally designed as a motor and has been employed as a generator without any modification.

Worked Example 4.2

A six-pole 50 Hz induction motor runs at 4% slip at a certain load. Calculate the synchronous speed, the rotor speed, the frequency of the rotor currents, the speed of the rotor RMF with respect to the rotor and the speed of the rotor RMF with respect to the stator.

Model answer

The synchronous speed from Equation (4.4) is $N_s = f/p = 50/3$ rev/s $= 50 \times 60/3 = 1000$ rev/min

The rotor speed from Equation (4.16) is $(1 - s)N_s = (1 - 0.04) \times 1000 = 960$ rev/s

The frequency of the rotor currents are: $f_r = sf = 0.04 \times 50 = 2$ Hz

The speed of the rotor RMF with respect to the rotor:

$$N_{rr} = f_r \times 60/p = 2 \times 60/3 = 40 \, \text{rev/min}$$

The speed of the rotor RMF with respect to the stator:

$$N_r + N_{rr} = 960 + 40 = 1000$$

(i.e. the rotor and stator RMFs rotate together)

4.4.2 The Induction Machine Equivalent Circuit

The induction machine can be viewed as a transformer with a rotating secondary. Imagine an induction machine with its rotor mechanically locked, i.e. at standstill. The stator RMF will be rotating at ω_s with respect to the rotor and inducing in each phase the voltage E_2 at mains frequency f. The current that flows in each phase will be

$$I_2 = \frac{E_2}{R_2 + j2\pi f L_2} = \frac{E_2}{R_2 + jX_2} \tag{4.19}$$

where R_2 and L_2 are the effective per-phase resistance and inductance of the rotor winding and X_2 is the rotor reactance at mains frequency. At standstill the slip $s = 1$ and the rotor voltages and currents are of the stator frequency f. At any other rotor speed, the slip is s, the

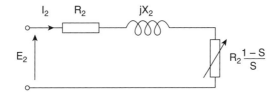

Figure 4.13 Equivalent circuit of induction machine rotor

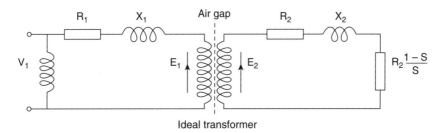

Figure 4.14 Induction machine stator-rotor equivalent circuit

induced voltage is sE_2, the rotor frequency is sf and the rotor reactance is sX_2. For the rotor current at slip s we can write the more general expression:

$$I_2 = \frac{sE_2}{R_2 + jsX_2} = \frac{E_2}{(R_2/s) + jX_2} = \frac{E_2}{R_2 + R_2[(1-s)/s] + jX_2} \tag{4.20}$$

since $R/s = R_2 + R_2[(1 - s)/s]$.

Equation (4.20) provides the rational for the equivalent circuit shown in Figure 4.13. This circuit resembles that of the secondary winding of the transformer (Figure 4.10) but with a variable resistive load connected to its output. The energy conservation principle indicates that the electrical power transferred to the rotor is the real part of $\boldsymbol{E_2 I_2^*}$. The power lost irreversibly in the rotor ohmic resistance is $R_2 I_2^2$ and the remainder, i.e. $I_2^2 R_2[(1-s)/s]$, must be and indeed is the electrical power converted into mechanical power.

The transformer equivalent circuit analogy is extended in Figure 4.14 to include the stator parameters. Here R_1 and X_1 represent the stator winding resistance and inductance while X_m represents the magnetizing reactance drawing the current necessary to establish the RMF. The dashed line corresponds to the air gap interface across which energy is transferred from the stator to the rotor.

In a further simplification the ideal transformer can be omitted by transferring elements from secondary to primary using the transformation ratio. In Figure 4.15 R_s and X_s are the stator winding resistance and reactance respectively. The elements R_r and X_r represent the rotor resistance and reactance respectively referred to the stator using the rotor–stator transformation ratio. The product $R_r[(1-s)/s]I_s^2 = R_{em}$ represents the electrical power per-phase converted into mechanical power. This equivalent circuit tells us that when the rotor is locked, $s = 1$, $R_r[(1 - s)/s] = 0$, so all the input to the rotor is converted into heat in R_r. When $s < 1$, the energy into the rotor is partly converted into heat in the winding resistances and partly into mechanical form.

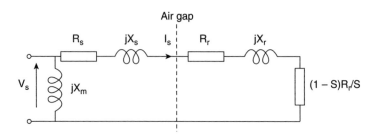

Figure 4.15   Induction machine equivalent circuit referred to the stator

With the induction machine generating, the slip is negative and the notional resistance $R_r[(1 - s)/s]$ is also negative. This is perfectly consistent with circuit analysis. A positive RI^2 implies irreversible conversion of electrical energy into thermal energy. A negative RI^2 implies the conversion of some other type of energy (in this case mechanical) into electrical.

Irrespective of whether the machine is motoring or generating the mains to which the machine is connected supplies the reactive voltamperes absorbed by all the inductive components of the equivalent circuit. This must be so as only positive or negative real power is associated with the mechanical/electrical energy conversion in the resistance $R_r(1 - s)/s$. The consequence is that induction generators always absorb reactive power from the mains.

4.4.3 The Induction Machine Efficiency

If the total electrical input power per phase fed into the stator is P_s, the power crossing the motor's air gap i.e. the power per phase transferred from the stator to the rotor is $P_r = P_s - R_s I_s^2$. All of P_r is dissipated in $\{R_r + R_r[(1-s)/s]\} = R_r/s$ so, $P_r = (R_r/s)I_r^2$. Hence the rotor copper loss is:

$$R_r I_r^2 = s P_r \tag{4.21}$$

Subtracting the rotor copper loss from P_r gives the average per-phase mechanical power $P_m = P_r - R_r I_r^2$ which through substitution from Equation (4.21) gives

$$P_m = (1 - s) P_r \tag{4.22}$$

The developed torque Q_m of the motor is its total mechanical power $3P_m$ divided by the motor shaft speed ω_r. Therefore

$$\dot{Q}_m = \frac{3 P_m}{\omega_r} \tag{4.23}$$

Substituting Equations (4.16) and (4.22) into (4.23) gives

$$Q_m = \frac{3(1 - s) P_r}{(1 - s)\omega_s} = \frac{P_{ag}}{\omega_s} \tag{4.24}$$

where $P_{ag} = 3P_r$ is the total three-phase power crossing the air gap.

Neglecting the stator copper losses and rotating mechanical losses, the efficiency of an induction motor is given approximately by:

$$\eta = \frac{P_{out}}{P_{in}} = \frac{P_m}{P_r} = \frac{(1-s)P_r}{P_r} = (1-s) \qquad (4.25)$$

For an induction generator (s negative) the power flow is in the reverse direction hence approximately

$$\eta = \frac{P_{out}}{P_{in}} = \frac{P_r}{P_m} = \frac{1}{(1-s)} \qquad (4.26)$$

Equations (4.25) and (4.26) indicate that for the conversion efficiency to be high, s at full load must be as small as possible.

Real induction generators have losses that have not been taken into account in this simplified analysis. The mechanical power available to produce electricity is reduced by windage and other mechanical frictional losses within the generator. Additionally, electrical and magnetic losses within the rotor reduce the power that is transferred from the rotor across the air gap to the stator. Finally, in the stator there are more losses associated with the winding resistance and the setting up of the magnetic excitation in the shunt branch of the equivalent circuit. As a consequence, large induction generators have efficiencies in the region of 90%. These extra losses will be referred to in a later section.

4.4.4 The Induction Machine Speed–Torque Characteristic

An important characteristic of any electromechanical converter is its speed–torque relationship. For the induction machine the developed torque Q_m from Equations (4.23), (4.22) and (4.16) is

$$Q_m = \frac{3P_m}{\omega_r} = \frac{3(1-s)P_r}{(1-s)\omega_s} = \frac{3P_r}{\omega_s}$$

Substituting for

$$P_r = \frac{R_r}{s}I_s^2 = \frac{R_r V_s^2}{sZ^2}$$

and using the equivalent circuit of Figure 4.15 we get

$$Q_m = 3\left(\frac{R_r}{s}\right)\frac{I_s^2}{\omega_s} = \frac{3R_r}{s\omega_s}\left[\frac{V_s^2}{(R_s + R_r/s)^2 + (X_s + X_r)^2}\right] \qquad (4.27)$$

This relationship is nonlinear and can be generalized to describe typical performances of an induction machine by normalizing it in terms of torque and speed. Taking as normal torque the rated torque and as normal speed the synchronous speed, the normalized relationship of Equation (4.27) for a typical induction generator is plotted in Figure 4.16 with R_r as a parameter.

Note that, as for the synchronous machine, there is a maximum or *pullout* torque beyond which the generator will accelerate uncontrollably. However this condition is far away from the normal operating regime. The curve for rotor resistance 'R_r' represents the performance of a typical induction machine with low rotor resistance and shows that the variation in speed from zero input torque to rated torque varies by about 3–4%. For s small, R_r/s is large compared to R_r and $X_s + X_r$ and to a good approximation Equation (4.27) can be written as

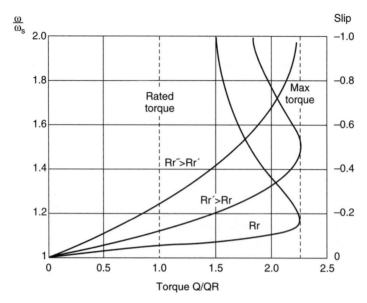

Figure 4.16 Normalized speed and slip against torque for a generator

$$Q_m \approx \frac{3V_s^2 s}{\omega_s R_r} \tag{4.28}$$

Equation (4.28) indicates that in the normal operating range (zero to rated torque), torque is directly proportional to slip and therefore speed and is inversely proportional to R_r. Curves for $R_{r'} > R_r$ and $R_{r''} > R_r$ are also plotted on Figure 4.16. By selecting the value of the rotor resistance, a designer has the ability to change the slope of the torque–speed characteristic. If a substantial variation of speed with torque is required, the rotor can be designed to have a large resistance. The downside of such an arrangement is the unacceptable reduction in efficiency.

A method to access the rotor windings and therefore exploit the property of speed change is to arrange a rotor that has coils rather than short circuited bars, with the coil terminals connected to slip rings and brushes so that additional external resistance can be connected in series with the windings. In such a *rotor wound* induction machine, the rotor winding is similar to that on the stator. The disadvantages of the wound rotor induction generator include a higher capital cost and a higher maintenance cost.

Worked Example 4.3

A wind turbine rated at 450 kW has the following induction generator parameters in ohms: $R_s = 0.01$, $X_s = X_r = 0.15$, $R_r = 0.01$ and $X_m = 6$. At a time when it is supplying its rated output the slip is 0.01. Calculate the mains voltage and the power factor at which the induction generator is supplying power to the grid using the simplified equivalent circuit.

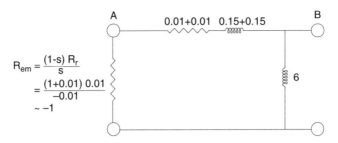

Figure 4.17   Equivalent circuit for Worked Example 4.3

Model answer

The circuit in Figure 4.17 brings to mind the synchronous generator equivalent circuit of Figure 4.7. The reactance to resistance ratio of the series impedance in figure 4.17 is 0.3/0.02=15; hence, to a good approximation, Equations (4.8) and (4.9) can be used. The mechanical power converted from the wind into electrical power per-phase is 450 000/3 = 150 000 W. This appears as a 'negative' power dissipation in R_{em} therefore the voltage V_A across R_{em} is given by

$$150\,000 = \frac{V_A^2}{R_{em}}$$

and

$$V_A = 387\,\text{V}$$

Using Equation 4.8(a)

$$150\,000 = \frac{V_A V_B}{X}\sin\delta = \frac{387 V_B}{0.3}\sin\delta$$

and

$$V_B \sin\delta = 116.28$$

The reactive power at end A is zero, hence, from Equation (4.8b).

$$0 = \frac{V_A}{X}(V_A - V_B\cos\delta) = \frac{387}{0.3}(387 - V_B\cos\delta)$$

and

$$V_B \cos\delta = 387$$

Hence

$$\tan\delta = \frac{116.28}{387} \quad \text{and} \quad \delta = 16.7°$$

giving $V_B = 404\,\text{V}$, i.e. $404\sqrt{3} \approx 700$ line volts

The reactive power associated with end B of the line is given by Equation (4.9b); hence

$$Q_B = \frac{V_B}{X}(V_B - V_A \cos\delta) = \frac{404}{0.3}(404 - 387\cos16.7) = 44874\,\text{VAR}$$

The Q absorbed by the shunt reactance $X_m = 6\,\Omega$ is given from

$$Q_m = \frac{404^2}{6} = 27202$$

The total Q per phase is $= 44\,874 + 27\,202 = 72\,076$. Hence the power factor is

$$\cos\tan^{-1} = \frac{72076}{150000} = 0.9$$

4.4.5 Induction Generator Reactive Power

Worked example 4.3 shows that the induction generator is a source of active power but a sink of reactive power. Even when the real power output from an induction generator is zero it will still draw considerable reactive power through X_m to magnetize its iron core (3 × 27.2 kVAR in the worked example). As increasing torque is applied and more real power is exported to the network, extra reactive power is absorbed due to the reactive power consumed by the series reactance (3 × 44.87 kVAR in the example). A typical relationship between active and reactive power for an induction generator is shown in Figure 4.18.

The induction generator power factor will vary from zero at A to around 0.9 at B. In order to improve the power factor it is often necessary to fit local power factor correction (PFC) capacitors at the generator terminals (Appendix). These have the effect of shifting the overall characteristic downwards to A′ B′. The amount of reactive power 'compensation' required depends on a number of technical and economic factors.

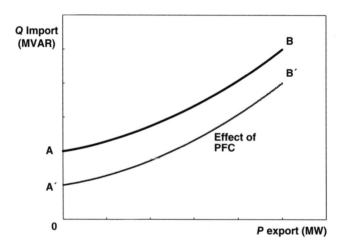

Figure 4.18   Relationship between active and reactive power for an induction generator

Table 4.2   Comparison between synchronous and asynchronous generators

Synchronous generator	Induction generator
Features	
• Efficient	• Moderately efficient
• Expensive	• Less expensive
• Requires maintenance	• Rugged and robust, little maintenance
• Reactive power flow can be controlled through field current	• Sink of reactive power
• Fixed speed hence very stiff	• Small change in speed with torque, hence more compliant
• Responds in an oscillatory manner to sudden changes in torque	• Responds to sudden torque inputs in a nonoscillatory way
• Can be built with permanent magnets for a large number of pole pairs and low rotational speed ('ring' form)	• Cannot be built economically for low rotational speeds
• Suitable for variable speed operation through a power electronic interface	• Suitable for variable speed in its rotor wound form in conjunction with a power electronic converter
• Suitable for connection to weak networks. Used in autonomous systems	• Suitable for week networks only in conjunction with power electronics
• Requires special synchronization equipment to connect to mains	• Can be simply synchronized to the mains
Use in RE generation	
• Used in wind power mainly in its 'ring' form for gearless coupling to a wind turbine	• Used extensively with a gear box in wind power
• Variable speed provided through a DC link power electronic interface	• Variable speed provided with power electronics in the rotor wound form
• Used in water power when reactive power control is required	• Used in water power with gearbox

4.4.6 Comparison between Synchronous and Asynchronous Generators

Table 4.2 provides an overall comparison of the characteristics of induction and synchronous generators with particular reference to renewable energy applications.

4.5 Power Electronics

4.5.1 Introduction

Power electronics is concerned with the application of electronic devices to control and condition electrical power. Power control involves the regulation of the power transfer from the renewable energy generator to the mains either to maximize this transfer continuously as the available resource changes or to limit the transfer for operational reasons. Power conditioning involves the transformation of power from one voltage/current/frequency/waveform to a different voltage/current/frequency/waveform.

A power electronic interface that controls and/or conditions power is referred to as a *converter*. Converters are interposed between the RE generator and the mains and often carry the total power transfer. They should be designed so that as little power as possible is lost in this transfer and that their capital cost is as slow as possible per watt transferred. Transformers can be viewed as power conditioning devices. Converters, not unlike transformers, are rated in volt-amps (VA).

The devices (transistors, etc.) used in power electronic converters are made from semiconductor silicon, like those in a computer processor, but physically they are much larger, and usually they are discrete – one device per piece of silicon. A converter will typically contain between one and a hundred power semiconductor devices.

Power electronic converters are found in an enormous range of applications and sizes, from the 2 GW UK-France Channel DC Link down to the 40 W energy-saving light bulbs in homes. In renewable energy, power electronic converters are already used in most PV systems and many wind turbines. It is also expected that they will be used in the future in practically all the emerging wave and tidal power technologies.

Despite the enormous range of converter sizes and applications, they are mostly based on just a handful of device types and basic circuit configurations. This section will discuss the most common of these. The aim is not that the reader will be able to design converters, but that he or she will understand their specifications, potential uses and limitations. The following section will deal with the application of these converters in the conditioning of power from renewable energy sources.

More details on solid state switching devices and converters can be found in many books on power electronics (see, for example, Reference. [3]). Some of the most common devices are reviewed below.

4.5.2 Power Semiconductor Devices

Diodes

The *diode* in Figure 4.19 conducts current from anode to cathode, (in the direction of the arrow) whenever the anode is made positive with respect to the cathode, i.e., positive *bias voltage* is applied across the device, and blocks current in the reverse direction when the device is negatively biased. The diode has no capability to control the current after it has been established. Diodes are readily available with ratings up to thousands of volts and thousands of amps. They are very reliable and very cheap (per VA) and very useful.

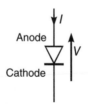

Figure 4.19 Diode

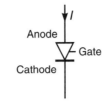

Figure 4.20 Thyristor

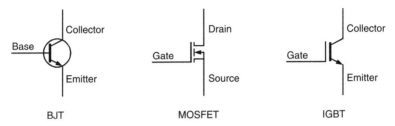

Figure 4.21 Transistor types

Thyristors

In contrast to the diode, the *thyristor* (Figure 4.20) has a control capability through a third electrode known as the *gate*. Even if the thyristor is positively biased no current will flow through the main anode–cathode circuit until a small *pulse* has been applied to the gate. Thereafter, the thyrstor acts like the diode so that the current will continue to flow through the anode–cathode circuit until that current reduces to zero. The thyristor itself cannot reduce or switch off the current. Unlike the diode, however, after the current has reached zero, the thyristor regains its capability to block current in the forward direction.

Thyristors were the first commercially available controllable power semiconductor devices and were used in converters of all sizes. They are still the cheapest controllable devices and are used in very high power converters (hundreds of MVA) and in smaller cost-sensitive applications.

Gate turn-off thyristors (GTOs) and integrated-gate commutated thyristors (IGCTs) are devices based on the thyristor principle but capable of turning themselves both on and off. These derivatives are expensive and are being superseded by advances in transistor technology.

Transistors

Bipolar junction transistors (BJTs), metal oxide semiconductor field effect transistors (MOSFETs), insulated gate bipolar transistors (IGBTs) (Figure 4.21) are all significantly more expensive than simple thyristors, but have the big advantage that they can be turned off by a control signal. Practical and economic considerations currently favour the use of MOSFETs for small converters up to about 30 kVA, which includes most PV inverters. IGBTs are used in converters up to about 10 MVA, which includes wind turbine applications.

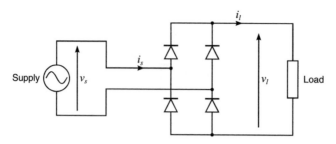

Figure 4.22   Circuit of a single-phase diode bridge rectifier

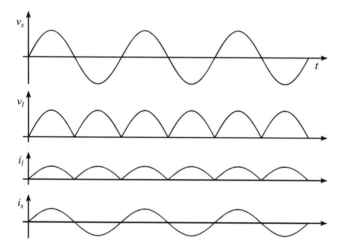

Figure 4.23   Waveforms in a diode bridge rectifier connected to a purely resistive load

4.5.3 Diode Bridge Rectifier

In Figure 4.22 alternating current power is fed from the supply to the diode junctions. It is rectified by the four diodes that form the bridge, and the resulting DC power is supplied to the load on the right. Thus, the circuit converts the AC power to DC. Assuming, for simplicity, that the load is purely resistive, the resulting waveforms are shown in Figure 4.23.

The voltage supplied to the load is DC, in that it is always positive, but it is far from smooth because it contains a considerable *ripple*. Very few practical loads would tolerate a supply like this except those, e.g. heating resistors, that would be equally happy on an AC supply. For this reason, it is common to add a smoothing capacitor as shown in Figure 4.24.

The waveforms in Figure 4.25 show that, as before, the diode bridge is rectifying AC power to DC. Now, however, the voltage supplied to the load v_l is almost constant; a bigger capacitor would reduce the ripple even further. The capacitor acts as an energy store so that when the output voltage from the rectifier is lower than the voltage across the capacitor, energy is supplied to the load from the capacitor. When the output voltage from the rectifier is larger than the capacitor voltage the capacitor is charged from the supply. This charging effect is provided

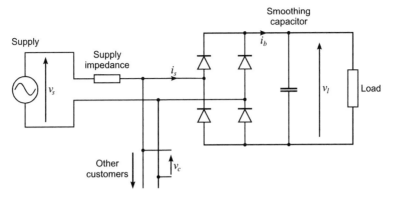

Figure 4.24   Circuit of a diode bridge rectifier with a smoothing capacitor and a supply impedance

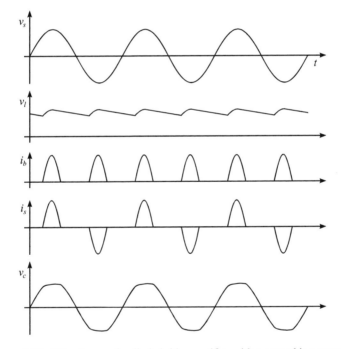

Figure 4.25   Waveforms of a diode bridge rectifier with a smoothing capacitor

by a series of pulses shown as i_b in the figure. Because the duration of the pulses is short, their amplitude must be high, if the same average power is to be supplied to the load as it was supplied before the capacitor insertion. The downside of this arrangement is that the current drawn from the AC source i_s is no longer a sine wave. Instead, it is a series of positive and negative pulses.

4.5.4 Harmonics

The nonsinusoidal current i_s in Figure 4.25 may cause problems for the AC supply network. In particular, transformers and cables in the network will experience additional heating. Despite this, the circuit shown in Figure 4.23 and variants with the same problem are very widely used, because they are cheap. TVs and computers are the biggest offenders, mainly because there are so many of them. Indeed, in areas where the load is dominated by TVs and computers, it is common for the customer's voltage waveform v_c in Figure 4.25 to be *flat-topped*, due to the nonsinusoidal voltage drop in the supply impedance. Plotting instantaneous supply current i_s against voltage v_c would show a highly nonlinear relationship. Appliances such as TVs and computers are sometimes called nonlinear loads.

Another way of describing the problem is by considering the harmonic content of the waveforms which can be obtained through Fourier analysis. For example, the supply current i_s in Figure 4.25 has a very large third harmonic component. This is a particular problem in three-phase systems (in just about every large power system in the world) because the third harmonics in the neutral conductor do not cancel (Appendix). Thus, the normal assumption that the neutral conductor carries zero current in a balanced three-phase system no longer holds. In general, all harmonic currents cause undesirable heating in the transformers and cables of the supply system. In a bid to prevent excessive harmonic currents in distribution networks, electricity utilities have instigated the introduction of regulatory standards that set maximum permissible levels for the harmonic currents that individual appliances may cause.

In general, all power electronic converters cause some harmonic currents in the AC network to which they are connected. Converters, used in large high voltage DC (HVDC) transmission systems are particularly prone to this. Indeed, much of the engineering design of such converters concentrates on reducing the harmonic currents to an acceptable level. It is therefore understandable that electricity utilities have expressed concern regarding the suggestion that, in future, large numbers of inverters will be used to connect PV and other renewable energy sources to distribution networks.

Fortunately, the converters used in modern renewable energy systems use techniques that reduce the low order harmonics to negligible levels. In the context of a network supplying a typical collection of TVs and computers, etc., the harmonic contribution from renewable energy systems can be expected to be negligible.

4.5.5 The Thyristor Bridge Converter

Replacing the diodes in the bridge of Figure 4.22 with thyristors, as shown in Figure 4.26, allows the DC voltage to be controlled. In many applications, the anode–cathode current is that of a line in an AC circuit, and thus, switching-off occurs when the line current naturally reaches zero. The transfer of current from one conducting device to another previously non-conducting device is called *commutation*. Thyristors are solely suited in converter applications where the commutation process is carried out by the AC supply. Such circuits are known as *line commutated* converters.

Note that the load now includes an inductance, which is typical of practical applications of thyristor bridges. This inductance is usually large enough to ensure that the current i_{dc} is

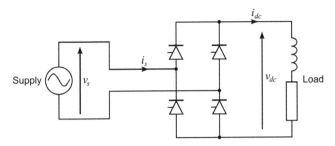

Figure 4.26   Circuit of a single-phase thyristor bridge

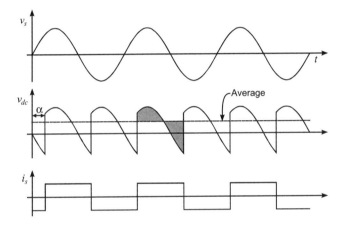

Figure 4.27   Waveforms in a thyristor bridge, while rectifying

nearly constant, at least in the short-term from one cycle to the next. In the longer term, it will vary to match the average level of v_{dc}.

A control circuit (not shown in Figure 4.26) applies firing pulses to the thyristor gate connections. The thyristors are fired in diagonal pairs. The fired pair take over the conduction of the current, which reduces the current in the other pair to zero and switches them off. Control of the firing angle α provides a means of controlling the average DC voltage. Figure 4.27 shows the situation when $\alpha = 50°$. The shaded areas, above and below the average DC voltage, are equal (because, in the steady state, the average voltage across the inductance must be zero). At $\alpha = 90°$, the average DC voltage would be reduced to zero. If α were increased past $90°$, the average DC voltage would theoretically become negative. This is an impossible operating condition in the circuit of Figure 4.26, because the direct current would be required to reverse direction, i.e. flow from cathode to anode through the thyristors. However, if the load resistor were to be replaced by a DC source, whose voltage was larger than the average negative voltage of the converter, the DC current direction will be maintained, power will be transferred from the DC source to the AC side and the bridge will operate as an *inverter*. Rotating the circuit diagram in Figure 4.26 through $180°$ puts the positive DC rail back at the top and gives the normal way of drawing an inverter, with the power flowing from left to right as in Figure 4.28.

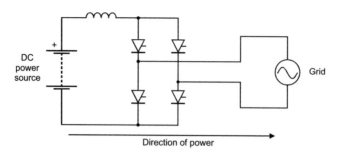

Figure 4.28   Line commutated thyristor inverter

Note that the inductance on the DC side is still required. It causes the direct current feeding the inverter bridge to be pretty much constant (from one cycle to the next), which leads to this circuit being called a *current-fed* or *current-source inverter*.

The thyristor inverter shown in Figure 4.28 has several shortcomings:

1. The current waveform on the AC side is almost a square-wave (see Figure 4.27) and therefore has very high harmonic content.
2. The fundamental component of the square-wave alternating current is out of phase and lagging the AC voltage. The circuit consumes reactive power from the grid.
3. The circuit requires connection to the grid (or a synchronous generator) in order for the currents to commutate. Therefore, the circuit cannot be used as a general purpose stand-alone AC supply.

Nonetheless, thyristor inverters and controlled rectifiers, based on the above concepts, were the foundation of the power electronics industry during the 1960s and 1970s. Thyristor invert-ers were used in early variable speed wind turbines and grid-connected PV systems. The shortcomings of the converter based on thyristors are all but removed by the use of transistor switching devices.

4.5.6 The Transistor Bridge

During the 1980s, self-commutated power semiconductor devices (BJTs, MOSFETs, IGBTs, GTOs and IGCTs) started to become commercially viable, which led to the development of *self-commutated inverters* an example of which is shown in Figure 4.29. In a self-commutated inverter, the switching, both on and off, of the main power semiconductor devices is under control. Thus, a self-commutated inverter does not need a grid connection in order to operate and can be used to create a standalone AC power supply, the frequency of which is determined by the transistor firing control circuit.

Note that the inductance on the DC side is no longer required. This inverter is fed with a voltage that is pretty much constant (from one cycle to the next), which leads to it being called a *voltage-fed* or *voltage-source inverter*. On the other hand, for reasons explained later, an inductance is required on the AC side if the inverter is to be connected to the grid.

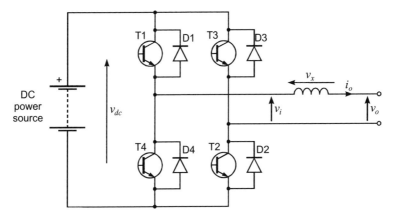

Figure 4.29 Transistor bridge

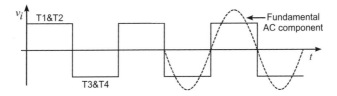

Figure 4.30 Voltage waveform of a basic square wave inverter

The diodes, shown next to each of the transistors, make the circuit appear more complicated but, in practice, this is not an issue. They are necessary in order to provide a path for the current when the transistors are turned off.

Basic Square Wave

Applying a very simple switching pattern to the transistors in the bridge circuit shown in Figure 4.29 allows it to produce a basic square wave as in Figure 4.30. A basic square wave can be used to supply some non-critical loads such as incandescent light bulbs and simple universal motors, but it is not suitable for the majority of AC loads, which are designed to operate from a sinusoidal supply.

The basic square wave has a sinusoidal component at the fundamental frequency, as shown in Figure 4.30, but it also has a very high harmonic content. These harmonics will typically cause excessive heating in induction motors and transformers. For the same reason, a basic square wave is not acceptable for grid connections.

Quasi-Sine Wave (Modified Square Wave)

A minor modification to the control system allows the bridge circuit to produce a quasi-sine wave. As shown in Figure 4.31, the quasi-sine wave is really just a modified square wave, but it is a much better approximation to a sine wave. Setting the switching angles to 60–120–

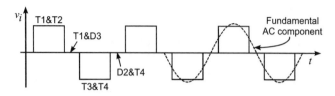

Figure 4.31   Voltage waveform of a quasi-sine wave inverter

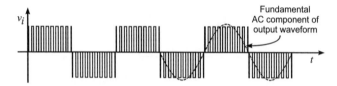

Figure 4.32   Pulse-width-modulated (PWM) inverter output voltage

60–120° eliminates the third harmonic completely and makes the waveform good enough for many practical loads. The switching angles may be adjusted slightly to control the RMS voltage, bearing in mind that this adjustment increases harmonics.

Pulse-width Modulation (PWM)

For loads that are more harmonic-critical and for grid-connection, *pulse-width modulation* (PWM) allows the transistor bridge circuit shown in Figure 4.29 to produce an almost-pure sine wave (low order harmonics are virtually eliminated) and provides full control of its amplitude. In a pulse-width-modulated (PWM) inverter, the transistors are switched at a much higher frequency than that of the intended output waveform. The width (duration) of the high frequency pulses, having short width at the edges and increasingly longer width towards the centre of the waveform, as shown in Figure 4.32, is controlled so as to create a good approximation of a sine wave output.

In order to provide a high quality sine wave, the switching frequency needs to be as high as possible. If it is too high, however, *switching losses* will become significant, making the inverter itself less efficient. Transistors when at their on or off state dissipate very little power. It is during the periods of transition between the two states that they are most lossy. For small MOSFET PWM inverters, switching frequencies up to 20 kHz are typical. While PWM virtually eliminates the low order harmonics, there can be significant harmonics around the switching frequency and its multiples, but these can readily be filtered out.

Figure 4.33 illustrates one simple method that the inverter's internal control system may use to create the switching signals. The pulses that switch the transistors on and off are generated at the intersections of the reference sine wave with the carrier wave which is usually triangular. In a standalone inverter, the reference sine wave would be created by the internal control system, normally to provide a constant voltage and constant frequency output. Reference [3] provides more information on this rather complex topic.

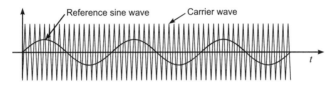

Figure 4.33   PWM construction of a switching pattern

Comparison of Switching Methods

Quasi-sine wave inverters are widely used for small standalone applications. For larger applications, and for grid-connected inverters, PWM is normally employed.

The above discussion has focused on inverters providing AC power at 50 or 60 Hz, but inverters are also used to provide power at much higher frequencies. This is very useful internally in converter systems because it allows the use of high frequency transformers, which, as mentioned earlier, are much smaller, lighter and cheaper than 50/60 Hz transformers of the same power rating. This is a topic that will be visited later when converters for renewable energy sources will be reviewed.

Output Control in a Grid-Connected Inverter

In a grid-connected inverter, the reference sine wave shown in Figure 4.33 is created so that it is of the same frequency as that of the grid but it can be phase-shifted with respect to the grid voltage. Referring to Figure 4.29 and using phasors rather than instantaneous values, it is clear that

$$V_i = V_o + V_x$$

where V_i is the fundamental component of the pulse-width modulated voltage output from the inverter bridge, V_o is the output voltage, which is the grid voltage once the inverter is grid-connected, and V_x is the voltage across the inductor. Furthermore, the voltage across the inductor must lead the output current I_o by 90°, ie:

$$V_x = jI_oX$$

where X is the inductor reactance.

Phasor diagram (a) in Figure 4.34 illustrates the general case and shows that control of the amplitude and phase of V_i with respect to the grid voltage V_o provides control of both amplitude and phase of the output current I_o. This is essentially the mechanism for the control of the active and reactive power flows as described in Equations (4.8a) and (4.8b) in the case of the synchronous generator.

Phasor diagram (b) shows the particular case where the inverter is operating at unity power factor (zero reactive power flow) where the output current I_o is in phase with the grid voltage V_o. The required amplitude of V_i and its phase with respect to the grid voltage can readily be calculated for any required output current I_o.

Furthermore, neglecting losses in the circuit, we know that power-out must equal power-in, i.e.:

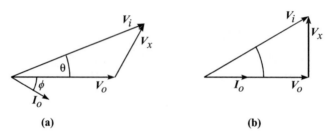

(a)   (b)

Figure 4.34   Phasor diagrams of the output stage of a grid-connected inverter; (a) general case and (b) at unity power factor

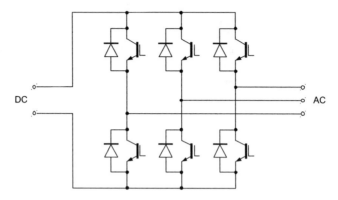

Figure 4.35   Three-phase IGBT bridge

$$V_{dc}I_{dc} = V_oI_o \cos\phi$$

where I_{dc} is the average value of the current drawn from the DC source. Thus, by using the PWM pattern to control the output, the input current has been controlled.

The Three-phase Bridge

All of the preceding circuits and discussions may readily be extended to three-phase systems. Usually this requires only an additional 50% of devices as shown in Figure 4.35 where a single-phase bridge with four transistors has becomes a three-phase bridge with six transistors. Inverters above 10 kVA are usually three-phase.

Three-phase power electronic converters are extensively used in variable speed wind turbines to convert AC to DC and vice versa. The circuit shown in Figure 4.35 is known variously as a transistor bridge, an inverter, a voltage-source converter, a variable frequency drive or just about any combination of those words. It is widely used throughout industry to operate induction motors at variable speed. IGBT converters are expensive, typically several times the cost of the associated electrical machine.

4.5.7 Converter Internal Control Systems

Very simple converters may employ analogue electronics to provide the transistor gate or base control signals. In all other cases, the internal control of a modern converter is provided by a dedicated microprocessor or digital signal processor (DSP). By controlling the exact timing of transistor gate or base control signals, the microprocessor has control over the currents, voltages and power flow through the converter. The microprocessor will often be used to provide a number of other functions. In a grid-connected PV inverter, for example, the microprocessor will typically perform the maximum power point tracking (MPPT), anti-islanding protection and perhaps some diagnostics and communications.

Depending on the application, the microprocessor based control system in a modern converter may have a variety of control objectives. For example, in a standalone inverter, the objectives would be to keep the output at constant voltage and constant frequency; in a grid-connected PV inverter, the objective would be to achieve maximum power extraction as the solar radiation varies, a task that would require control of the inverter output current; and in a motor drive inverter, the objective might be to control the torque. The power electronic circuits used in all three of these examples could be very similar, but the implemented control systems make them behave very differently.

4.5.8 DC–DC Converters

An AC voltage can be easily stepped up or down through the use of the ubiquitous transformer. Unfortunately, this facility was not as easily available for direct current until the recent explosion in power electronic developments. As will be shown later, a DC transformer widens considerably the application of power electronic converters.

An ideal DC–DC transformer will convert power at one voltage to power at a different voltage. If the voltage is increased, the current will be decreased by the same factor, and vice versa, such that:

$$V_{in}I_{in} = V_{out}I_{out} \qquad (4.29)$$

DC transformers are known as *choppers.* or *DC–DC converters.* In such choppers a power transistor is switched at high frequency and the duty cycle (the on-to-off time ratio known as the mark-space ratio) controls the output voltage.

Step-down DC–DC Converter

Provided that the current in the inductor is continuous, the steady state operation of the circuit in Figure 4.36 can be described as follows. The transistor is switched on for time T_{on}, during which $v_1 = v_{in} - v_{out}$. Then it is switched off for a time T_{off}, during which $v_1 = -v_{out}$. In the steady state, the volt-seconds across the inductor must be zero, so

$$V_{out} = DV_{in} \quad \text{where } D \text{ is the duty cycle } \frac{T_{on}}{T_{on} + T_{off}} \qquad (4.30)$$

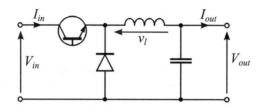

Figure 4.36 Step-down DC–DC Converter (also known as a series switching regulator or buck converter)

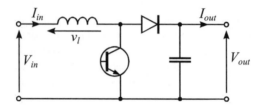

Figure 4.37 Step-up DC–DC converter (also known as a boost converter)

Step-up DC–DC Converter

Provided that the current in the inductor is continuous the steady state operation of the circuit in Figure 4.37 is given by

$$V_{\text{out}} = \frac{1}{1-D}V_{\text{in}} \qquad\qquad (4.31)$$

4.6 Applications to Renewable Energy Generators

4.6.1 Applications to PV Systems

PV System Characteristics

All PV systems generate direct current and so power electronic converters, known as inverters, are required to connect them to power systems.

As described in Chapter 2, PV modules have an I–V characteristic that depends on incident radiation intensity (and to a lesser extent the solar spectrum) and module temperature. Different types of PV have different shapes of I–V curve, characterized in simple terms by a fill factor, as explained in Chapter 2. To maximise the output of any PV module, or collection of modules, operation should be at the maximum power point. This point will change with radiation and temperature and a key function of the inverter is to track this. How effective this so-called maximum power point tracking (MPPT) is can critically affect the overall efficiency of the PV system.

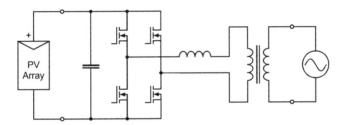

Figure 4.38   Basic grid-connected PV inverter

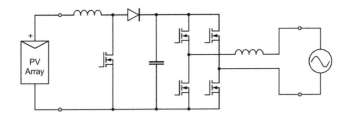

Figure 4.39   Transformerless grid-connected PV inverter

Basic Grid-connected PV Inverter

The grid-connected PV inverter shown in Figure 4.38 is based on the transistor bridge circuit of Figure 4.29. It is required that the DC voltage of a PV array is inverted and connected to a grid of, say, 230 V. For such a simple, e.g. domestic application, the array voltage would have to be well over 350 V DC at the maximum power point, which would be impractical in many cases. The transformer is therefore required in order to raise the output inverter voltage.

The control could be implemented as described earlier under the heading 'Output control in a grid-connected inverter' in Section 4.5.6. Control of the AC output current would provide control of the DC current drawn from the array and hence allow maximum power point tracking. The main drawback of the circuit is the size, weight and cost of the output 50 or 60 Hz transformer.

Transformerless Grid-connected PV Inverter

The addition of a DC booster stage, at the input of the inverter, as shown in Figure 4.39, removes the need for the mains frequency transformer, and reduces the current in the inverter transistors as they operate at a higher voltage than in Figure 4.38. The converter in Figure 4.39 has two stages, both of which require to be controlled.

One possibility would be to operate the DC booster at a fixed duty cycle, which would give a fixed ratio between the average DC currents and allow the maximum power point tracking to be implemented by controlling the AC output current as it was in the previous example. In practice, it may be more efficient to fix the voltage of the DC bus (the DC con-

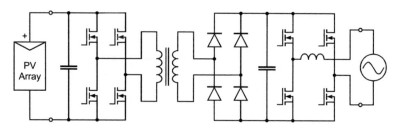

Figure 4.40   PV Inverter using a high-frequency transformer

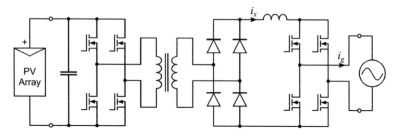

Figure 4.41   PV Inverter using a steering bridge

nection from the booster to the inverter). In this case, the control of the inverter current is used to provide control of the DC bus voltage – if the voltage is too high, more current is drawn, and vice versa. Maximum power point tracking is then implemented by controlling the duty cycle of the DC booster.

A perceived drawback of this converter is that it provides no electrical isolation between the grid and the PV array. This means that the array is at mains voltage and that for safety requirements it must be double insulated. Some commercial inverters currently being manufactured are based on this arrangement.

PV Inverter Using a High Frequency Transformer

The inverter shown in Figure 4.40 has three stages. Starting from the left, the first is a high frequency inverter, which could be a quasi-square-wave inverter, or a square wave inverter. Operation at 20 kHz is typical and this allows the transformer that follows to be much smaller, lighter and cheaper than the mains transformer used in Figure 4.38. The transformer steps up the voltage and provides the isolation that was missing in the circuit of Figure 4.39. Next, the high frequency AC is rectified to provide DC suitable for the final stage PWM inverter on the right of the diagram. Although this circuit appears complicated, the cost reduction of the transformer makes it attractive to manufacturers.

PV Inverter Using a Steering Bridge

The inverter shown in Figure 4.41 again makes use of a high frequency transformer, but differs from that of Figure 4.40 in that the power flow through the transformer is now modulated at 50 or 60 Hz. This modulation is achieved by use of a more sophisticated switching pattern on the input (left-hand) inverter and is designed to make the current i_x take the shape of a rectified 50 or 60 Hz sine wave. The output (right-hand) bridge then 'un-rectifies' or 'steers' this to produce a full sine wave current into the grid. The advantage of this, over the arrangement shown previously in Figure 4.40, is that the steering bridge switches at just 100 or 120 Hz, and only when the instantaneous current is zero, making it more efficient than the PWM inverter on the right of Figure 4.40.

4.6.2 Applications to Wind Power

Fixed Versus Variable Speed – Energy Capture [4]

This section deals with average changes of wind speed over tens of seconds or minutes and the wind turbine response to maximize energy capture over this timescale. In Section 2.4.3 mention was made of the fact that for maximum energy extraction the wind turbine speed should be adjusted to vary in sympathy with the wind speed. Figure 4.42 shows typical wind turbine power versus rotational speed characteristics with wind speed U as a parameter where $U_1 < U_2 < U_{rated}$. For each curve the converted power is zero at standstill ($\omega = 0$) and at runaway speed when the turbine is rotating without any restraining torque. In between these two extremes the power rises to a maximum. It stands to reason that for each wind speed it is desirable to operate the wind turbine at this maximum power point. For this to be possible as the average wind changes, the turbine torque must be adjusted so that the locus aa that passes through the maximum power points is tracked. This is an alternative way of expressing

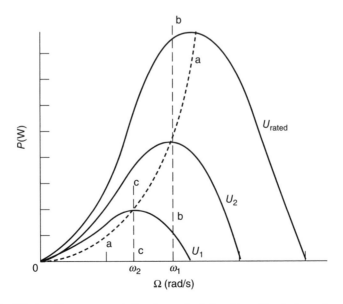

Figure 4.42 Wind turbine power–rotational speed relationship with wind speed as a parameter

the requirement stated in Section 2.4.3 where the tip speed ratio $\lambda = \Omega R/U$ should be fixed at λ_{opt}, the value at which C_p is a maximum. For each wind turbine design λ_{opt} is different, e.g. in Figure 2.8, C_p is a maximum when $\lambda \approx 7$.

Unfortunately, synchronous generators and, to all intents and purposes, cage rotor asynchronous generators, operate at a fixed speed, say at ω_1 in Figure 4.42, and a turbine driving either generator would track a locus such as bb. Only at wind speed U_2 would the wind turbine operate at λ_{opt} with the consequence that for higher and particularly lower wind speeds less than maximum energy would be extracted. Operation at variable speed (to track locus aa) with either synchronous or asynchronous generators is possible through the use of power electronic converters. Due to the losses of such converters the additional energy captured by a wind turbine operating at variable rather than fixed speed is typically less than 10% and would not alone justify the extra expense. However, there are other reasons why variable speed operation is desirable.

Fixed Versus Variable Speed – Dynamics

The vast majority of the world's electricity is generated by synchronous machines directly connected to their respective power systems. This configuration works very well when the prime mover (usually an engine or a steam, gas or water turbine) provides a steady torque to the generator. Wind, however, is turbulent and this translates directly into fluctuations in drivetrain torque. Using a directly connected synchronous generator in a wind turbine would form too rigid a coupling between the mechanical and electrical systems. Wind gusts would cause large mechanical stresses in the turbine and, depending on the nature of the electrical grid, large fluctuations in the power fed into the electrical system.

The main advantages of variable speed wind turbines in terms of dynamics are:

- The total inertia of the aerodynamic rotor, the gearbox (if there is one) and the electrical generator act as an energy buffer, smoothing out the wind turbulence. Transient torques and rapid variation in electrical power as well as stresses in the drivetrain are greatly reduced
- Lower structural loads and lighter foundations are other advantages of particular importance in offshore applications

Additional advantages are:

- The power electronics may also be capable of regulating the reactive power flow in the network.
- Noise is reduced, when operating at low wind speeds.

The drawbacks of variable speed are the extra complexity of the generator (in some schemes) and of the necessary power electronic hardware, all of which increase cost and possibly reduce reliability.

To allow for wind gusts, virtually all wind turbines have generators or generator systems that provide some degree of speed variation.

Synchronous Generator Supplying an Autonomous Network

A synchronous generator driven by a turbine through a gearbox is exclusively used to supply an autonomous network. In such networks the wind turbine may be the only source of power or more often may be supported by a storage system, e.g. a battery plus converter, or another source such as a diesel driven generator. At present, the relatively few examples of this scheme are restricted to small islands or to remote communities. If the wind turbine is the main source of energy, it is also in charge of system frequency and voltage. For a fixed frequency grid, the wind turbine has to be controlled to operate at fixed speed in spite of variations in wind speed and in system demand. This can be achieved through (a) a stall regulated wind turbine with frequency sensitive prioritized load switching or (b) a combination of variable pitch wind turbine control plus a controllable dump load and/or a storage system such as batteries with a bidirectional converter. Such autonomous systems are at present unusual and will not be referred to further.

Fixed Speed Wind Turbines

Figure 4.43 shows the classical turbine–gearbox–asynchronous generator–grid system used extensively in the 1980s. It is simple, reliable and very well proven in practice. Many of the wind turbines installed during the 1980s and 1990s were of this type and are still in service today. In this scheme the induction machine is of cage configuration and, due to the inevitable slip, the speed varies marginally with power so the speed is not strictly constant. The small mechanical compliance due to the slip provides some degree of electrical power smoothing during gusty conditions and is of benefit to reducing loads in the mechanical drivetrain.

Where a large induction machine is to be directly connected to the network (without speed control), some form of *soft starter* is generally required in order to limit the currents and torque during start-up. A soft starter is not a converter as such but simply limits the surge of current necessary for an asynchronous generator to absorb in order to establish the rotating magnetic field when first connected to the mains. As it uses thyristors it is much cheaper than the transistor based interfaces.

Figure 4.44 shows a power–electronic soft starter comprising three pairs of antiparallel thyristors. Initially, the thyristors are fired very late in the cycle so that little current flows. Gradually, the firing is advanced until the thyristors are effectively 'on' continuously, at which

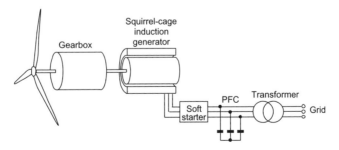

Figure 4.43   Directly connected squirrel-cage induction generator

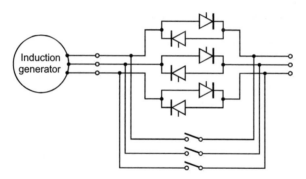

Figure 4.44 Soft starter

time the contactor at the bottom of the diagram can be closed and the soft starter's job is complete. The induction generator is usually a 4-pole machine, which therefore has a shaft speed close to 1500 rpm on a 50 Hz system (1800 rpm on 60 Hz). Thus, a gearbox is required.

Following the soft starter there is a set of power factor correction (PFC) capacitors. The induction generator naturally consumes reactive power in setting up its rotating magnetic field. This reactive power could be imported from the grid, but the network operator would normally charge for it, making it more economic to create locally. A bank of capacitors is usually installed in the nacelle or base of the tower to bring the overall power factor close to unity through the power factor correction principle discussed in the Appendix. Sometimes the capacitors are divided into sections that can be independently switched so that the reactive power created can be adjusted to match that consumed by the induction generator as the operating conditions change.

Lastly, Figure 4.43 shows a transformer that may be in the nacelle or in the base of the tower and may or may not be included in the turbine manufacturer's scope of supply. The induction generators in wind turbines usually operate at 690 V and the transformer steps this up to the voltage of the wind farm power-collection network (typically 11 kV, or 33 kV in the UK). Large wind farms may have a further transformer to step up the voltage again before connection to the utility grid.

Two-speed Wind Turbines

The operating speed of an induction machine is based on the synchronous speed, which is determined by the number of poles, which is in turn determined by the arrangement of the stator windings. The main generators used in wind turbines often have four poles, which gives a synchronous speed of 1500 rev/min on a 50 Hz grid. Use of six or eight poles gives 1000 and 750 rev/min respectively. It is possible to change the number of poles in an induction machine by switching the connections to individual stator windings or by the use of two generators. The low speed generator, operating at speed ω_2 in Figure 4.42, is used to improve energy capture and reduce noise when the average wind speed is low. Since the power available from the wind is also low, the rating of this generator is between one-quarter and one-sixth of the main generator. When it is detected that the average wind is above a predetermined level, the low speed generator is disconnected from the mains and the high speed generator takes over and operates at speed ω_1. Several Danish manufacturers used this scheme extensively in the 1980s and 1990s.

Variable Slip Wind Turbines

The squirrel-cage rotor, used in the fixed speed wind turbines described above, is very economic, robust and reliable. It has a small but finite resistance, which largely determines the slope of the torque–speed characteristic and hence the magnitude of the slip. Larger squirrel-cage induction machines generally have proportionally lower rotor resistance and therefore less slip. This reduces the compliance between the mechanical and electrical systems, which is not desirable in wind turbines. Figure 4.16 illustrates one way in which the speed of the induction machine could be regulated. The rated torque can be developed at different speeds by increasing the rotor resistance from R_r through to $R_{r'}$ and then to $R_{r''}$. This can be easily achieved by introducing an external set of variable resistors across the slip rings of a rotor wound induction machine. Controlling the speed in this way, however, incurs an unacceptable reduction in efficiency due to the energy dissipated in the external resistors. This mode of control could be dismissed as useless in wind turbines, nevertheless, a manufacturer has used it intelligently and successfully to improve performance.

In a wind turbine, the resistance can be controlled dynamically to allow for wind gusts. During the gust, the resistance is increased, allowing a little more (negative) slip; the rotational speed increases slightly and energy is absorbed by the inertia. After the gust, the resistance is reduced, which brings the speed back down and exports the energy to the grid over a longer period of time. Figure 4.45 shows such a scheme. If this scheme were to be applied to wind speeds between cut-in and rated, valuable energy from the wind would have been irreversibly dissipated in the external resistors. However, if this technique is applied at times when the average wind speed is above rated i.e. when energy in the wind can be shed without economic penalties, an interesting benefit results. In what follows, the rational of this policy will be explained.

In the Vestas OptiSlip scheme, [5, 6], a specially designed asynchronous generator with a wound rotor is fitted with rotating integrated electronics and resistors on the shaft, thus eliminating the need for slip-rings and brushes. Here the blade pitch control mechanism takes

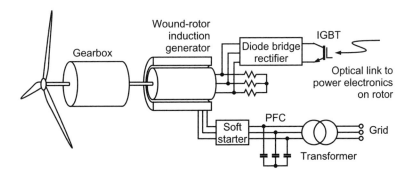

Figure 4.45   Vestas OptiSlip rotor resistance controller

care of the longer variations in the wind, while the OptiSlip system looks after the faster variations due to gusting. The power output is maintained at the rated value with the speed allowed to vary between the 1 and 10% slip range. The variable slip concept produces high quality electrical output at high wind speeds without generating any detectable harmonics. The variable slip concept does not enhance energy capture but improves the quality of generated electricity and the stresses in the wind turbine drivetrain in the region that matters, i.e. for wind speeds above rated.

The Principle of Slip Energy Recovery

A far superior approach to dissipating energy in external rotor resistors is to capture this energy and transfer it to the mains. There are a variety of such *slip energy recovery* schemes. In this section the general principles of slip energy recovery will be outlined before dealing with its application to wind turbines. At the outset it must be remembered that the frequency and magnitude of the induced voltages and the resulting currents in the rotor circuit of an asynchronous generator depends on the slip. Transferring energy out of or into the rotor circuit requires the interfacing of this variable frequency variable magnitude rotor voltage to the fixed frequency, fixed magnitude voltage of the mains. This can be achieved through an *asynchronous link* consisting of an AC to DC converter followed by a DC to AC converter.

The technical problems in such a scheme are not trivial. At very low slips the rotor side converter will have to deal with very low voltage, low frequency rotor voltages. Conversely, at high slips, the voltage and frequency will be proportionately larger. It can be shown that for control of speed from zero to synchronous, the rating of the rotor connected DC link will have to be the same as the nameplate rating of the induction generator. This would involve an electronic interface of considerable size and therefore cost. The advantages offered by full variable speed control can be almost matched if a limited range of control is provided. This can be ingeniously arranged by a rotor connected asynchronous link of relatively low rating. To investigate this arrangement, the possible power flows in and out of the rotor of an asynchronous machine need to be looked at more closely.

Figure 4.46(a) shows the power flow in an induction machine when motoring. The electromagnetic power P_{ag} crossing the air gap is equal to the electrical power P_s fed into the stator minus the losses referred to in Section 4.4. In turn, subtracting the rotor losses from P_{ag} gives the useful mechanical power output P_m. Neglecting the losses, the simplified diagram of Figure 4.46(b) can be set up with the aid of Equation (4.22). The asynchronous machine behaviour when the machine is generating will now be examined, and with the assumption that the capability exists of extracting and injecting power into the rotor circuit through an asynchronous link. In other words, there is an ability to adjust the magnitude and sign, i.e. the direction of P_r in Figure 4.46(b). In the figure, the arrows indicating power flows are shown in the conventional way from left to right, i.e. when the machine is motoring.

In the usual operation as a generator (Figure 4.16) the slip is negative in the range $-1 < s < 0$, and P_m flows from right to left, i.e. has a negative sign. As a consequence P_{ag} and P_s are both negative but P_r is positive. By extracting the required level of P_r from the rotor circuit, the induction generator speed can be controlled in the *supersynchronous* range. In this mode, the mechanical input power is fed into the mains partly through the stator and partly through the rotor connected asynchronous link.

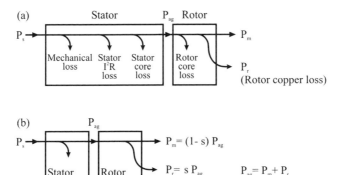

Figure 4.46   Power flows in induction machine

Through the asynchronous link, the capability exists of injecting rather than extracting power into the rotor circuit, thus reversing the sign of P_r. The consequence of this is that as P_m and P_{ag} are negative the slip now must be positive and in the range $1 > s > 0$. The induction generator can therefore be made to operate in a *subsynchronous* speed. In this mode of operation $P_m < P_{ag}$ and hence the electrical power fed into the rotor from the mains is returned to the mains through the stator. Due to the possible bidirectional flow in the rotor circuit this machine is often referred to as being *doubly fed*.

It can be shown that the rating of the asynchronous link in the rotor circuit is proportional to the required speed shift from synchronous; for example a speed shift of 30% will require an asynchronous link rated approximately at 30% of the nameplate rating of the induction generator. The advantage of modifying the slip energy should now be obvious. By having a bidirectional link capable of extracting or injecting 30% of the rated power in the rotor circuit, a 60% speed variation can be achieved. For the purposes of reaping the advantages of variable speed operation this is almost as good as full speed control. Also, because only 30% or less of the power is going through the converters, energy losses in the converters and harmonic generation are reduced. It is not surprising that a number of major manufacturers of wind turbines have chosen this technique for their multimegawatt wind turbines.

DFIG Wind Turbines [4]

DFIG stands for a *double fed induction generator*. Figure 4.47 shows a simplified schematic of a wind turbine connected to a doubly fed induction generator. The rotor-connected voltage source IGBT converter is linked to a mains-connected IGBT converter through a DC link containing a capacitor. Both converters are capable of *four-quadrant* operation; i.e. they can be controlled to transfer active and reactive power in either direction. The rotor-connected converter has a more onerous task to perform than the mains-connected one.

The rotor-connected converter looks after the wind turbine speed control so that energy extraction is maximized. The generator is controlled on the basis of *vector control*, a complex technique that requires substantial on-line computational requirements. This technique 'decouples' the active and reactive power control functions permitting the induction generator to be controlled with exceptional flexibility. It also offers superior transient response permit-

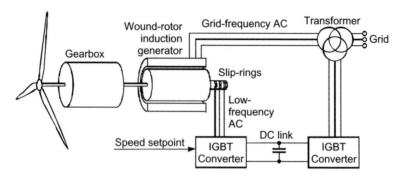

Figure 4.47 DFIG (doubly fed induction generator)

ting the induction generator to follow commands on torque change very rapidly. This enhances the energy extraction capability from gusts and relieves drivetrain stresses. For this computation to be executed, it is necessary to feed into the converter controller information on the rotor speed and its angular position, the magnitude and phase of the rotor and stator currents and information on the torque–speed curve that must be followed to maximize energy extraction.

Reactive power, or power factor, could also be externally specified. Although either converter can be arranged to inject reactive power into the mains (the former through the rotor–stator path, the latter directly into the mains), it was found that the task can be more economically performed through the rotor-connected converter.

The function of the mains-connected converter is, in most schemes, to transfer the required active power from the DC link to the grid or vice versa. This is carried out through a voltage control loop which ensures that the DC voltage is maintained fixed. For example, if the rotor-connected converter decides that more energy should be transferred from the wind turbine to the DC link (as a response to an increase in wind speed), the DC link capacitor will be overcharged and its voltage will increase. This will prompt the mains-connected converter to transfer more energy from the DC link to the mains, restoring voltage equilibrium. Because of the substantial variation in the magnitude of the voltage in the rotor of the asynchronous generator, the DC rail voltage is kept low. For an economic design this necessitates the connection of the mains-connected converter to the grid through a transformer.

Both converters are controlled through PWM techniques which ensure minimum harmonic injection into the mains. To protect the power electronics, an overvoltage protection is mounted on the rotor terminals to short circuit the rotor in case of equipment malfunction. The converters in Figure 4.47 are rated to carry only the rotor power, typically 30% of the wind turbine rating.

Technical details of the various wind turbine generator concepts presented above can be found in reference [8].

Wind Turbines with Full Converters

The configuration shown in Figure 4.48 uses two full converters. Full, in this context, means that all the power goes through the converters. The converters allow the standard squirrel-cage

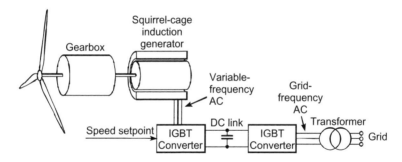

Figure 4.48   Wind turbine with an induction generator and full converters

induction machine to operate at variable frequency and therefore 0 to 100% variable speed. The active power from the induction generator is rectified by the machine-side converter and fed to the DC link. The grid-side converter acts as an inverter, taking power from the DC link and feeding it to the grid.

The structure of this link is very similar to that of Figure 4.47. Vector control of the induction generator is again implemented. The reactive power demand of the induction generator is supplied by the rotor-connected converter, whilst the mains-connected converter has the task not only of transferring to the mains the whole of the active power generated but also of injecting or absorbing reactive power to/from the mains respectively. Few manufacturers have adopted this option, which requires a cheaper machine but a more expensive DC link compared to the DFIG.

The converters used in this scheme are standard industrial drives, as would normally be used for variable speed motors; they are reliable and readily available from numerous manufacturers. The main drawback of the configuration is that it requires two IGBT converters rated at the full power of the wind turbine and these are expensive. The benefit is that the generator is a very robust squirrel-cage machine and the DC link is more system friendly during AC network faults and disturbances.

Synchronous Generators in Wind Turbines

As noted earlier, the main reason for using an induction generator in a wind turbine is that for fixed speed the slip provides some compliance and nonoscillatory response (Table 4.2), and for variable speed, the ingenious rotor power recovery scheme provides economic speed control. However, in the case of a fully variable speed wind turbine a synchronous generator may again be considered.

In this context, a synchronous generator has two advantages. Unlike an induction generator it does not need a supply of reactive power and therefore the output can be connected to a DC bus through a simple bridge rectifier using diodes or thyristors. It can be manufactured with a very large diameter and large number of poles. Such a machine has a low operating speed and high torque, and can be designed for direct mechanical coupling to the rotor of a wind turbine, eliminating the need for a gearbox.

A synchronous generator can be constructed either with a wound rotor carrying a controllable DC field excitation current or with a permanent magnet rotor. Both approaches have been demonstrated in large scale wind turbines. A wound rotor has the advantage of control-lability of the generated voltage through field current control. This reduces the DC link duties compared to a rotor with permanent magnets in which the generator voltage is just proportional to the speed. However, with permanent magnets the generator is more efficient due to the absence of the excitation loss. In either case the variable frequency AC in the machine stator need not be sinusoidal and the freedom to use a trapezoidal waveform, for example, can be exploited in the electromagnetic design of the machine. Additionally, as the AC output is rectified, it is not required to generate a frequency as high as 50 Hz at the highest rotational speed.

Gearless Wind Turbines

Use of a synchronous generator through a full converter also reopens the possibility of eliminating the gearbox. This approach has been successfully followed by the company Enercon since 1993. The first consideration in designing a gearless wind turbine is that the rated torque of any electrical machine is roughly proportional to the volume of its rotor. The gearbox in a typical 1 MW wind turbine has a ratio of about 70. Eliminating the gearbox increases the torque and hence the volume of the required generator rotor by roughly the same factor. To achieve the volume without the weight and to achieve a sufficient rate of flux cutting, a very large-diameter ring-shaped generator is used. It is still considerably heavier than the usual 1500 rev/min generator found in most wind turbines, but Enercon consider that this is justified by the elimination of the gearbox.

The Enercon ring generators have a large number of poles: 72 in the case of the 1.5 MW E-66. The rotational speed of between 10 and 22 rev/min thus produces a frequency of between 6 and 13.2 Hz. Also, their generators are six-phase, which helps to reduce the ripple on the DC coming out of the diode bridge rectifier. The gearless configuration has been used in all Enercon models from the 300 kW E-30 in 1993 to the new 4.5 MW E-112.

As shown in Figure 4.49, the machine-side converter could be a simple diode rectifier, which is typical in small wind turbines. In larger wind turbines, a controlled rectifier is useful. Depending on the choices above, the DC bus may or may not keep a constant voltage as the speed varies. The grid-side converter is likely to be an IGBT based inverter that is required

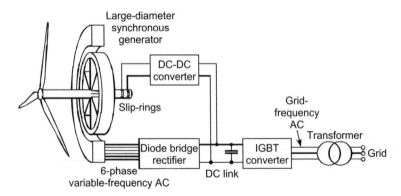

Figure 4.49   Gearless wind turbine using a synchronous generator

to transfer active power unidirectionally but is capable of injecting or absorbing reactive power from the grid.

At the time of writing a new generation of wind turbines based on what is known as the 'Multibrid' principle are being developed. These are a halfway house between the traditional systems that use high gearing and the more recent offerings that employ gearless coupling. The 'Multibrid' machines use a gearbox of moderate gearing driving a synchronous generator of moderate diameter, the whole system being packaged in a compact unit. Only time will tell whether direct drive schemes or 'Multibrids' will eventually supplant the well entrenched technology of variable slip induction generators.

Technical details of the various wind turbine generator concepts presented in this section can be found in References [8–10].

References

[1] Smith, R.J. and Dorf, R.C. Circuits, *Devices and Systems*, 5th edition, John Wiley & Sons, Ltd, Chichester.

[2] Gross, C.A. *Power System Analysis*, John Wiley & Sons, Ltd, Chichester, 1986.

[3] Lander, C.W. *Power Electronics* McGraw-Hill, Maidenhead, 1987.

[4] Slootweg, H. and De Vries, E. 'Inside wind turbines: fixed vs variable speed', *Renewable Energy World*, January–February 31–40.

[5] Pedersen, T.K. 'Semi-variable speed operation – a compromise?' Vestas-Danish Wind Technology A/S, Denmark, EWEA 1995 European Conference.

[6] Kruger and Andersen, B. 'Vestas Optispeed – advanced control strategy for variable speed wind turbines', Department for Research and Development, Vestas Wind Systems A/S, Jacobsens Vej 7, DK-6950 Ringkobing.

[7] Ekanayake, J., Holsworth, L. and Jenkins, N. 'Control of DFIG wind turbines' *IEE Power Engineering Journal*, February 2003.

[8] Heier, S. *Grid Integration of Wind Energy Conversion Systems*, 2nd edition, John Wiley & Sons, Ltd, Chichester, 2006.

[9] Ackerman, T. 'Transmission systems for off-shore wind farms', *Renewable Energy world*, July–August 2002, 49–61.

[10] Danish Wind Industry Association website www.windpower.dk.

5

Power System Analysis

5.1 Introduction

The preceding chapters have discussed electrical generators in some detail. Attention will now turn to the transmission and distribution systems that carry the generated power to the loads. In addition to their obvious task of carrying electricity over a distance and allowing generators and loads to be geographically separated, transmission and distribution systems also serve to provide interconnection. The interconnection of loads provides valuable aggregation and hence smoothing, as discussed in Chapter 3. Also, the parallel operation of generators allows inefficient part-loaded operation of individual generators to be largely avoided (Chapter 7), and thus overall efficiency is greatly increased. Furthermore, the parallel operation of generators and of sections within the transmission and distribution systems provides redundancy, allowing the overall impact of any single component failure to be minimized. Thus, even in situations where generators are located close to loads, the benefits of an interconnected power system should not be underestimated.

The effective operation of a power system requires that the routing of power from generators to consumers is carried out through the network so that transmission equipment is not overloaded, losses in the network are kept to a suitably low level, voltages at various nodes are maintained within limits and loss of generation or transmission equipment does not lead to an unacceptably high frequency of blackouts. To ensure that power systems operate securely and efficiently, special computational techniques have been developed by utilities. This chapter briefly describes these techniques and illustrates their use in determining the voltage at the nodes of the network, the complex power flows in transmission lines and the currents during fault conditions. The effect on the network of adding renewable energy generators can be assessed using such techniques, as will be examined in Chapter 6.

5.2 The Transmission System [1]

Figure 1.12 in Chapter 1 gave a pictorial view of a typical power system and the desirability of using various voltage levels for transmission were explained. Overhead lines, supported

Figure 5.1   Overhead lines at four different voltage levels. (Courtesy of Murray Thomson)

on steel towers (pylons), concrete towers or wooden poles, are highly visible in most power systems worldwide. Figure 5.1 shows the typical appearances of overhead lines at various voltage levels in the UK. It should be noted, however, that pole and tower constructions vary considerably and do not always provide a clear indication of the voltage level in use. Note also that underground cables, despite being much more expensive to install, are widely used at the lower voltage levels to avoid overhead transmission line intrusion in areas of habitation.

The 11 kV overhead line on wooden poles in Figure 5.1 is three-phase with no neutral. This is called a *single circuit*. The steel tower lines are carrying two three-phase circuits. In the 400 kV line each of the six conductors is made up of a bundle of four individual conductors. Each bundle behaves effectively as a single conductor, but the spacing helps to reduce the line impedance and the losses and increases the line capacity slightly. The two three-phase circuits can be operated independently, which provides operational flexibility and greatly increases reliability. Such an arrangement is known as a *double circuit*. Again there is no neutral: the wire visible at the very top of the tower is an earth wire for lightning protection.

The voltages indicated in Figure 5.1 are widely used in the UK. Other countries use slightly different voltages, often with fewer intermediate levels, but the structure is always similar. The terms *low voltage* (LV), *medium voltage* (MV), *high voltage* (HV) and *extra high voltage* (EHV) are often used but are avoided here because definitions vary considerably from country to country.

Transformers, located in substations (Figure 5.2) or on poles, are used to transfer power between lines and cables operating at different nominal voltages. A mature network may have tens of thousands of transformers. Fortunately, transformers are both very reliable and efficient.

Figure 5.2 Transformer: 132 to 33 kV. (Courtesy of Murray Thomson)

Figure 5.3 One-line diagram

Lastly, and by no means least, a great deal of switchgear is installed as suggested in Figure 1.14. This allows sections of the system to be isolated for operational and maintenance purposes, or on the occurrence of faults.

5.2.1 Single-phase Representation

The Appendix explains in detail the reason why three-phase is exclusively used for the generation and transmission of electrical power. It is also shown that as long as the three phases are balanced, calculations can be performed by considering just one of the phases. Thus, it is common to draw just one phase in circuit diagrams. To achieve this, delta-connected components are represented by their equivalent star and a neutral conductor may be drawn in, even when there is no neutral in reality.

To simplify drawings even further, it is common to omit the neutral return and represent a three-phase generator supplying a load as in the *one-line diagram* of Figure 5.3. This one-line shorthand is universally used by power system engineers both to represent schematically and perform calculations in balanced power systems.

A one-line diagram of a power system is shown in Figure 5.4. The different voltage levels found in a large power system may be thought of as layers, each one feeding the one below it by way of transformers.

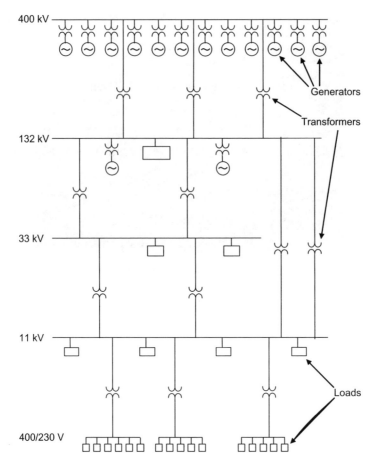

400 kV

Generators

Transformers

132 kV

33 kV

11 kV

Loads

400/230 V

Figure 5.4   One-line diagram showing 'voltage layers' in a large power system

5.2.2 *Transmission and Distribution Systems*

It is usual to refer to the higher voltage levels as the *transmission system* and the lower voltage levels as the *distribution system*, but the exact usage varies from country to country. In England and Wales, the 400 kV (plus some sections at 275 kV) is regarded as the transmission system, while the 132 kV and all layers below are the distribution system. In Scotland, many sections of 132 kV are regarded as transmission.

The term *national grid* usually means the transmission network. However, the term *grid* is often used loosely to describe the transmission and distribution networks. In particular, *grid-connected* means connected to any part of the network, but is used increasingly to mean small generators connected directly to the low voltage distribution system.

Nearly all generation that is currently installed is connected to the transmission system and is called *centralized generation*. Thus, traditionally, power always flows to the consumers down through the distribution system. Traditionally, distribution systems are designed with this in mind.

5.2.3 Example Networks

The transmission system shown in Figure 5.5 covers the length and breadth of the UK mainland. It is a single AC power system and therefore has the same frequency throughout. There is a similar, but smaller, transmission system that covers Ireland, including Northern Ireland. It also operates at 50 Hz nominal, but the frequency is not tied to the UK mainland system. There is a high voltage DC (HVDC) link from Scotland to Northern Ireland which allows power to be transferred in either direction, but does not require that the two systems are run in synchronism. There is a similar, but higher capacity, HVDC link between England and France. France is interconnected with its neighbouring countries so effectively there is a continuous connection between the most northern tip of Scotland and the most southern tip of Italy.

Most major European countries generate electricity approximately equal on average to their national demand. Nonetheless, interconnections with neighbouring countries are very valuable for efficient operation. Of particular note is Denmark, where the impressive penetration of wind power is facilitated by such interconnections.

Returning to the British Isles, many of the islands around the coast are now connected by under-sea cables and enjoy electricity with a cost and reliability similar to that on the mainland. More remote islands have standalone systems, usually with a heavy dependence on diesel fuel.

Figure 5.6 shows in greater magnification the north Wales network that includes only the upper voltage layers. The 400 kV transmission network matches (roughly) that shown in Figure 5.5. Transformers transfer power from the 400 kV to the 132 kV network which in turn supplies the major towns. Further transformer substations feed power down to the 33 kV network, which takes power to the smaller towns. Figure 5.7 shows the much denser 11 kV network layer which takes power to villages, hamlets and even individual farms. Finally, the even denser 400/230 V network (not illustrated) distributes power to individual properties, as shown schematically in Figure 1.12.

Further information regarding the UK power system can be found in Reference [1].

5.3 Voltage Control

Consumer loads comprise mainly heating and lighting elements, motors, electronic equipment providing audiovideo services, computers and controllers, battery charging facilities for portable equipment and finally electrochemical services mostly for industrial applications. All these 'loads' are designed to provide their service from a nominally fixed voltage supply. Utilities are therefore obliged by law to provide electricity at consumer terminals at a voltage that does not deviate from the nominal value by more than a few percent with a maximum of ±10% not being uncommon. For this to be achieved at the extremities of the distribution network the voltage of all nodes moving up through the layers of increasing voltage in the distribution and transmission system should be kept close to their nominal value. As the

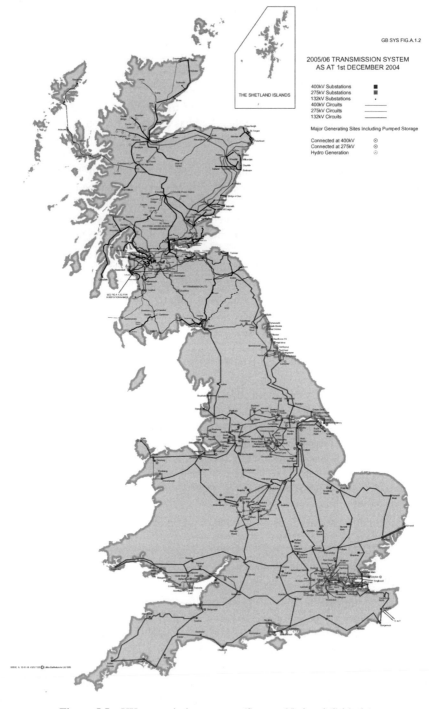

Figure 5.5 UK transmission system. (Source: National Grid plc)

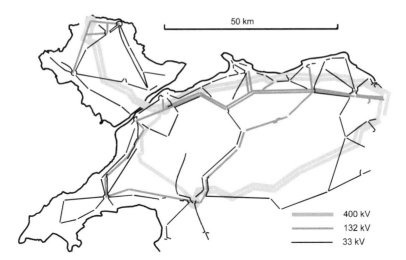

Figure 5.6   North Wales

Figure 5.7   11 kV in South Wales on the same scale

power system engineers would put it: 'the voltage profile from the generator terminals down to the consumer terminals should be kept nearly flat'.

The maintenance of a nearly flat voltage profile is disturbed by the fact that current flow through the impedance of transmission lines, cables and transformers causes unavoidable voltage drops. For these drops to be kept within permissible limits a number of *compensation* techniques have been developed over the years. These techniques have been considerably extended recently through the use of power electronic devices capable of high voltage and current ratings. Before discussing these techniques it is necessary to examine the characteristics of transmissions lines and the effect on voltage drops of the active and reactive power flows over the network.

5.4 Power Flow in an Individual Section of Line

Calculations to quantify power flows and exact node voltages throughout a network must take into account the network impedances. Later it will be shown that because of the operational complexities of power systems these calculations are not trivial and require sophisticated software for their execution. To gain an appreciation of these calculations and the relationships between power flows and network voltages, it is necessary first to consider a single point-to-point section of the transmission or distribution line.

5.4.1 Electrical Characteristics of Lines and Cables

The conductors used in overhead lines and underground cables are almost universally copper or aluminium, having a small but finite series resistance. They also have self and mutual inductance and shunt capacitance to each other and to ground (not shown), as illustrated in Figure 5.8. The resistance, inductance and capacitance are distributed along the length of the line and are quantified per unit length. For most calculations, however, they can be considered as lumped components with values calculated for the length of the line in question.

5.4.2 Single-phase Equivalent Circuit

Assuming that its operation is balanced, the three-phase line shown in Figure 5.8 can be reduced to a single-phase equivalent as shown in Figure 5.9. The capacitance has been removed in order to simplify the following calculations. Its effect is often negligible for short overhead lines. However, for long HV lines and invariably for underground cables, even of

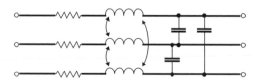

Figure 5.8   Resistance, inductance and capacitance of a three-phase overhead line or underground cable

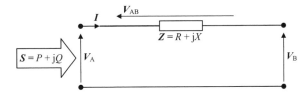

Figure 5.9   Single-phase equivalent circuit of a transmission line

short length, the capacitance is substantial. For such cases, the circuit of Figure 5.9 can still be used for the first stage of the analysis. The capacitance can be lumped at the two ends of the line and its effect taken into account at a second stage in the calculation.

In Figure 5.9, the upper line (containing the impedance) is one of the three lines shown previously in Figure 5.8. The lower line is a neutral, drawn in to assist visualization even though there is no neutral in reality. It is not known whether the components at the ends of the line are star or delta connected, but the calculations are being done as if both a star and a neutral were present.

P is the active (real) power flowing into the left end of the line and Q is the associated reactive power, defined in the same direction. They are jointly represented by the complex power S:

$$S = P + jQ \tag{5.1}$$

In this case, S, P and Q are per-phase quantities and will be one-third of the overall values carried by the real three-phase line. V_A is the complex phase voltage (line-to-neutral) at the left end of the line. It is equal to the line voltage (line-to-line) divided by $\sqrt{3}$. I is the complex line current and is the same as the phase current in this context. The complex power is related to the voltage and current by (Appendix)

$$S = V_A I^* \tag{5.2}$$

where I^* is the complex conjugate of I.

5.4.3 Voltage Drop Calculation

In Section 4.2.5 the synchronous machine equivalent circuit consisting of a single series reactance was used to express the relationships between the active and reactive power flows as functions of magnitudes and angles of the equivalent circuit terminal voltages (note the similarity between Figures 5.9 and 4.7). In what follows the inverse relationship will be derived because often there is interest in the way the voltage profile of the network from node to node is affected by the active and reactive power flows.

The complex impedance Z shown in Figure 5.9 includes resistance R and reactance X. The resistance is the same as that shown previously in Figure 5.8. The reactance is calculated from the inductance shown in Figure 5.8 plus mutual inductance terms. Such calculations are beyond the terms of reference of this book and can be found in any textbook on power systems, for example Reference [2].

Figure 5.10   Phasor diagram illustrating voltages and current shown in Figure 5.9

There will be a voltage across this impedance, according to Ohm's law:

$$V_{AB} = IZ \tag{5.3}$$

and, by Kirchoff's law,

$$V_B = V_A - V_{AB} \tag{5.4}$$

The phasor diagram of Figure 5.10 represents pictorially all the variables of Figure 5.9. Here the current I at end A is shown lagging the voltage V_A by about $30°$. Hence the transmission line at A is a sink of active ($+P$) and reactive ($+Q$) power and

$$S = P + jQ$$

Combining Equations (5.2) and (5.3) and taking V_A as the reference phasor with an angle of zero gives

$$
\begin{aligned}
V_{AB} &= \frac{S^*Z}{V_A} \\
&= \frac{(P - jQ)(R + jX)}{V_A} \\
&= \frac{PR + QX}{V_A} + j\frac{PX - QR}{V_A}
\end{aligned}
\tag{5.5}
$$

5.4.4 Simplifications and Conclusions

The use of complex numbers, as exemplified above, provides a systematic and accurate means of calculating magnitudes and phase angles of all quantities. Good engineering calculators and many software packages can manipulate complex numbers directly and make such calculations straightforward.

Nonetheless, the following approximations to Equation (5.5) are often very helpful, both for rough calculations and to get a better feel for the interdependencies. First, there is often interest in the scalar difference between the magnitudes of voltages V_A and V_B. Referring to Figure 5.10, it can be seen that this difference is contributed mainly by the real part of V_{AB} (its horizontal component). Therefore Equation (5.5) can be simplified to give a scalar relationship:

$$\Delta V \approx \frac{PR + QX}{V} \tag{5.6}$$

where $\Delta V = |V_A| - |V_B|$. The denominator V should really be $|V_A|$, but $|V_B|$ can be used, provided that ΔV is small. Equation (5.6) provides a convenient way to estimate the voltage

Table 5.1 Transmission line parameters

kV	Typical X/R ratio
400	16
275	10
132	6
33	2
11	1.5

at one end of the line compared to the other, for a given active and reactive power transfer over the line.

Moreover, the same equation can be used to estimate the voltage change that will occur at the point of common coupling (PCC) when a new generator (or load) is connected to a system. In this case, R and X need to represent the Thévenin equivalent impedance (Appendix) of the system as a whole, but more on this later.

A second approximation is applicable in transmission networks and the higher layers of distribution where, due to the geometry and size of the conductors, the impendence Z is dominated by the reactance X. Table 5.1 shows typical X/R ratios for high to low voltage overhead transmission lines. When the X/R ratio is high, $R = 0$ in Equation (5.6) with little loss in accuracy. This gives the very useful rule of thumb:

$$\Delta V \propto Q \tag{5.7}$$

In Figure 5.10 it can be seen that the angular displacement between V_A and V_B is to a large extent proportional to the imaginary component of the V_{AB} phasor. The imaginary part of Equation (5.5) gives a second rule of thumb:

$$\delta \propto P \tag{5.8}$$

where δ is the load angle between V_A and V_B, the same angle that was encountered in Chapter 4 when the operation of the synchronous generator was under study. Equations (5.8) and (5.7) match Equations (4.8a) and (4.8b).

These two relationships provide an important insight into the flow of complex power in transmission networks where $X \gg R$:

• Network voltages are determined largely by reactive power flow. Turning round this statement it can be affirmed that voltage magnitude differences between the ends of a transmission line is the primary driver of reactive power flow.
• Phase angles are determined largely by active power flow and are largely independent of reactive power flows. Turning round this statement as well it can be affirmed that angular differences between the voltages at adjacent nodes is the primary driver of active power flow.

These insights can be extended further. Consider that a power system component, capable of absorbing or injecting active or reactive power, is connected to a node of the transmission network. If the component were to inject or extract active power at the node, the node voltage angle will shift to a more leading or lagging position respectively with respect to the voltages of the adjacent nodes. If the component were to inject or extract reactive power at the node, the node voltage magnitude will increase or decrease respectively compared to its original value. It may be concluded that if a flat network voltage profile needs to be maintained, the high voltage can be depressed or the low voltage augmented at a node by connecting inductors or capacitors to that node respectively.

5.5 Reactive Power Management

The principle of conservation of reactive power described in Section A.13 requires that reactive power is scheduled so that as the consumer demand of Q varies over a day, appropriate Q resources are available to provide it. In Chapter 3 it was explained that a discrepancy between demand and supply of active power is manifested by rises or falls in frequency, a parameter that is common to the whole interconnected power system. In contrast, reactive power deficits or excesses manifest themselves as local system voltage drops or rises respectively. An overall deficit of reactive power in a system will manifest itself as unacceptably drooping voltage profiles. The reverse is the case for a surfeit of reactive power.

Ideally, reactive power should not be transported but generated locally where it is required. This, of course, is the principle of power factor correction of consumer loads, an arrangement described in Section A.14 and encouraged by utilities through tariffs. However, the vast majority of domestic and commercial loads as well as some industrial loads are not power factor corrected. Hence substantial amounts of reactive power have to be transported over the system and/or generated locally by special utility equipment. The picture is complicated by the fact that:

(a) High voltage overhead transmission lines overall generate Q when they are lightly loaded due to their shunt capacitance, but absorb Q when heavily loaded due to their series inductive reactance (Figure 5.8).
(b) Low voltage transmission lines have small shunt capacitance and overall are consumers of Q.
(c) Cables overall generate Q at all loads because of their substantial shunt capacitance.
(d) Transformers always absorb Q as they consist basically of two inductively coupled coils (Section 4.4.2).

5.5.1 Reactive Power Compensation Equipment

Utilities use the following devices to balance the Q flows in the network and therefore control the bus voltages:

(a) transformer tap changers;
(b) synchronous generator AVRs;
(c) shunt capacitors;
(d) shunt reactors;

(e) static VAR compensators (SVCs);
(f) fast AC transmission systems (FACTS);
(g) renewable energy generator interfaces.

(a) Tap Changers and Voltage Regulators

Tap changers play a major role in voltage control. A range of connection points (taps) to the windings of a transformer allows the turns ratio to be altered and thus provides voltage control. An on-load tap changer is a motorized switch allowing adjustment while the transformer is in service.

Tap changers do not themselves enable the generation or consumption of Q. Their function is to regulate the reactive power flow over the network from generators to consumers. An example of how this is done is described in the next subsection.

Figure 5.11 shows that on-load tap changers (transformer symbol with an arrow) are widely used throughout power systems, except on the 11 kV to 400 V final distribution transformers. The transformer symbol that does not consist of two linked circles represents an 'autotransformer'. This is a more economical transformer that consists of only one winding acting both as primary and secondary, often used to link some of the high voltage levels. There are a great many distribution transformers and the maintenance of motorized tap changers would be impractical. Instead, the taps on a distribution transformer can be adjusted manually, but this requires taking the transformer out of service.

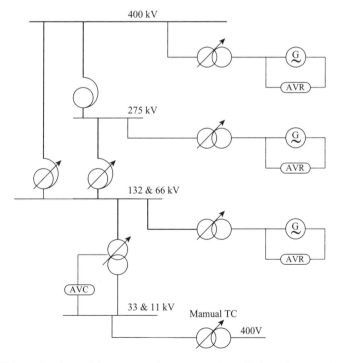

Figure 5.11 Voltage levels and interconnections to regulate Q flow in a typical transmission network

Thus, delivery of the correct voltage to domestic customers (230 V ±10%) relies heavily on the tap changer at the primary substation that feeds the 11 kV network. Because of its crucial importance this tap changer is activated through an automatic voltage controller (AVC). The allowable voltage range, ±10%, may seem generous, but in practice it is often fully exploited, to allow for load variation between summer and winter.

(b) AVRs

The internal source voltage in the stator of a synchronous generator (Figure 4.7) can be controlled by controlling the DC current (the excitation) in the field winding on the rotor. This control is often achieved by an automatic voltage regulator (AVR). This is an electronic controller that monitors the voltage at the synchronous generator terminals, compares it to a set value and if there is an error increases or decreases the excitation to nullify it. In the case of a standalone synchronous generator, the AVR's role is straightforward: it monitors and controls the terminal voltage.

Large conventionally fuelled generator units are connected to the grid through a 'unit' transformer fitted with a tap changer. Most auxiliaries that supply services to the generator are connected to the generator terminals and require a fixed voltage to provide the appropriate service. The task of the AVR here is to maintain the generator terminal voltage constant.

The level of reactive power injected to the grid by a generator connected to the grid through a transmission line is controlled by the unit transformer tap changer as follows. By increasing the nominal transformer ratio of, say, 22 kV : 400 kV by tapping-up, the transformer secondary voltage increases with respect to the remote end voltage of the transmission line and more reactive power is transported down the line. This extra reactive power flow through the synchronous reactance of the generator results in a reduction of its terminal voltage. This is sensed by the AVR which increases the excitation of the generator (hence the injection of extra reactive power) to bring the voltage up to the nominal level.

Generators fed from RE sources are substantially smaller in rating and are, in general, connected to the distribution rather than the transmission network. The effect on the system of such generators is discussed in Chapter 6. However, if the RE source is truly large, for example a large offshore wind farm or a tidal scheme, it could have an impact on the network. In such cases, more conventional control schemes may have to be used. In fact, as penetration levels by wind power increase, for example in Denmark where levels of 20% are no longer uncommon, utilities require renewable generators to be designed to contribute positively to power system voltage control.

(c), (d), (e) Static Compensators

At the transmission network, switchable capacitors and/or inductors are connected at crucial nodes of the system to regulate the voltage under varying loading conditions. These devices are controlled in steps and in the case of inductors are particularly lossy. For finer control, static VAR compensators (SVCs) are used. These employ solid state power electronic devices rather than mechanical switches to provide infinitely variable control of Q from reactors and capacitors.

(f) FACTS

The way PWM converters can be controlled to act as capacitors or inductors was discussed in Section 4.5.6. FACTS are power electronic converters capable of adjusting the angle and magnitude of the voltage of a network mode. They are capable therefore of regulating the active and/or the reactive power flow at a mode.

(g) RE Generator Interfaces

In Chapter 4 it was seen that in order to maximize effectiveness or because of absolute necessity the majority of renewable energy generators are interfaced to the grid through a power conditioning converter. This has the added advantage that if the converter is of the PWM type it can contribute to the voltage regulation at the PCC.

5.6 Load Flow and Power System Simulation [2, 3]

5.6.1 Uses of Load Flow

Section 5.4 looked at the maths describing an individual overhead line or underground cable. Mature power systems are likely to have thousands of such lines, all interconnected. The same basic maths applies to each and every line, but now the equations must be solved simultaneously. Structured procedures for such calculations are known as *load flow*.

A basic load flow calculation provides information about the voltages and currents and complex power flows throughout a network, at a particular point in time, with a given set of load and generation conditions. Additional information, such as losses or line loadings, can then be easily calculated.

Load flow analysis is an essential tool that provides the following vital information for the design as well as the operation and control of power systems:

- checking whether equipment run within their rated capacity;
- checking that voltages throughout the network are kept within acceptable limits;
- ensuring that the power system is run as efficiently as possible;
- ensuring that the protection system will act appropriately under fault conditions and that under likely contingencies the system will remain secure and operational;
- assisting in the planning of the expansion of conventional and renewable generation and the necessary strengthening of the transmission and distribution system to meet future increases in power demand.

Load flow calculations are central to the numerous software packages available for power system simulation, including PSS/E and DIgSILENT. Most such packages also include many additional functions, such as the following:

- *Fault analysis* determines the currents that will flow in the event of a short circuit. Such information is critical in the design of switchgear and protection systems.
- *Unbalanced analysis* allows modelling of networks where the three phases are not perfectly balanced. This is particularly relevant in networks where single-phase loads are connected.

- *Dynamic simulation* allows transients and stability to be examined. This is particularly important in the design of transmission systems. The modelling has to include the dynamic characteristics of the generators and their associated control systems.

5.6.2 A Particular Case

The one-line diagram of the Icelandic national power system shown in Figure 5.12 is powered mostly by renewable energy, primarily hydro, but with an increasing contribution from geothermal. The generators are indicated by the ⊘ symbol, while the loads are shown as triangles. In practice, these loads are the bulk supply substations that feed power to the distribution networks, but for now they will be thought of as simple point loads. The thick horizontal lines are *nodes* (also known as *buses*) and, although they have a finite length in the diagram, electrically they are just connection points: they have zero length and zero impedance. The thinner lines in the diagram are the actual transmission lines and each one has a finite impedance, which may be estimated from the line characteristics and its length. The system uses a mix of 220, 132 and 66 kV and the transformers connecting these are indicated by the usual ⊖ symbol.

In the design and operation of a power system it is important to know the power flows, both active and reactive, the voltages (which will normally differ slightly from the nominal

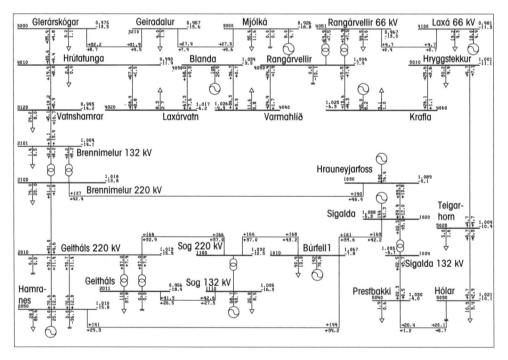

Figure 5.12   Single-line diagram of the Icelandic national power system. (Reproduced from Egill B. Hreinsson, www.hi.is/~egill/rit/estonia99.pdf)

values mentioned above) and the currents, throughout the system. The almost illegible numbers in Figure 5.12 are the values of these variables, which have been calculated by load flow analysis. How such calculations are performed will now be considered.

Consider the transmission line running across the middle of the diagram from node *Brennimelur* to node *Hrauneyjarfoss*. The power transfer over this transmission line will obey the mathematical analysis presented earlier in this unit. Indeed, the power transfer over every individual transmission line in the whole system must obey the equations presented earlier. The challenge is that, in general, the equations require to be solved simultaneously.

While the mathematical techniques and the software packages for performing load flow analysis are readily available, obtaining suitable input data can be much more difficult. The required input data can be divided into network data and load/generation data.

5.6.3 Network Data

The impedances of the individual lines are usually calculated from knowledge of the line type and the length of the line. In practice, particularly in lower voltage networks, these are not always known with absolute certainly. The line may have been installed fifty years ago and could have been modified since. Utilities do not always have perfect records. Also, particularly with underground cables, the details of the installation can significantly affect the characteristics; for example the dampness of the ground and the proximity of other cables can have a significant effect.

The ratios and impedances of the transformers must also be known and, again, records can lack details. Any tap-changing mechanisms on transformers must also be included in the model.

5.6.4 Load/Generation Data

Time Dependence

One load flow analysis solution provides the results for one operational instant. If any of the loads or generator outputs are changed, then the analysis must be run again. With modern computers, re-running the analysis is trivial; the challenge is in deciding what data should be used for loads and generators, given that they are continuously varying.

A common approach is to identify the worst cases, and typical choices are the winter maximum peak demand and the summer minimum, except in the US where the maximum is due to summer air-conditioning. At the winter peak, the system is stretched to its limit and it is necessary to ensure that no equipment is overloaded or any node voltage is below the minimum permitted. During the summer minimum, because of the low loading of transmission lines and cables, there may be an excessive reactive power available that would manifest itself as high voltage at nodes.

In a transmission system, the loads and generation normally vary smoothly and the relevant maximum and minimum figures can readily be identified. In lower voltage networks, the loads are much more variable and the worst case conditions can be very difficult to identify.

Types of Nodes (Buses)

PQ Buses

The obvious way of describing the load at a node is by specifying it as an impedance. Simple uncontrolled loads, such as heaters and incandescent light bulbs, tend to behave as near-constant impedances and thus the power they draw varies with the square of the voltage: a $\pm10\%$ variation in voltage, which is not uncommon, gives powers ranging from 81 to 121%. Many loads, however, include some sort of control mechanism such that the power they draw does remain near constant irrespective of the voltage changes. Most electronic based devices behave in this way.

Another important example is a distribution network that is supplied via a tap-changing transformer. Such a transformer senses the voltage on the lower voltage side and changes the taps so that this is maintained at nearly nominal value. A fixed load impedance on the transformer secondary appears on the primary side as a fixed P and Q demand irrespective of the transformer ratio. As the taps change the voltage on the primary side of the transformer may change substantially, but the P and Q are invariant. For example the tap-changing transformer at the primary substation can be adjusted to achieve a constant voltage on the 11 kV side irrespective of (modest) voltage changes on the 33 kV side. Thus, the power drawn by the 11 kV network from the 33 kV network is also held near constant. If a load flow calculation was being performed on the 33 kV network, which would typically feed several 11 kV networks, it would be reasonable to represent each primary substation as a constant P and Q. The same is true in load flow modelling of the higher voltage networks.

To summarize, assuming that the voltage at a load bus is kept constant, the load can be expressed as a fixed P and Q demand; such nodes are referred to as *PQ buses*. As will be explained later, this is a very convenient formulation for the purposes of carrying out the load flow.

Small renewable energy generators also fall in the *PQ* bus category. Distributed generators are infrequently called upon to control the network voltage. Instead, they are often configured to operate at near-unity power factor ($Q = 0$); in this case, it may be appropriate to label the node to which they are connected as a *PQ node*. In the case of fixed speed wind turbines, however, the reactive power consumed by the induction generator will be dependent on voltage, as will the reactive power generated by the power factor correction capacitors. Many load flow software packages include facilities to model induction machines and related equipment appropriately. The situation is similar with small hydro systems interfaced to the grid through induction generators. Energy from photovoltaic, wave and tidal schemes and MW sized wind turbines is fed to the grid through a power electronic converter. This provides the facility of reactive power injection/extraction at the point of connection.

To summarize, for relatively small embedded RE generators the P injection depends solely on the RE source (wind, sun, water) level at the time and the Q injection either on the bus voltage or on the setting of the power electronic converter. In the latter case the converter could be regulated to inject active power at a chosen power factor.

According to the generally accepted convention summarized in Figure A.20, at *PQ* nodes generators inject active power and so P is positive, whereas for loads, P is negative. The reactive power Q direction is defined similarly.

PV Buses

For large synchronous generators, the Q is often not specified, and instead it is the voltage V that is known. This is because such generators are fitted with AVRs that hold the V constant. To accommodate this, in load flow analysis nodes where such generators are connected are referred to as *PV buses* and are dealt with a little differently in the maths. Unfortunately, *PV* buses are sometimes described as *generator buses*, which makes sense so long as all the generators are large synchronous generators with AVRs.

As renewable energy generators increase in size, utilities have been developing regulations requiring that such generators behave in a traditional manner. Multimegawatt wind turbines connected to the network through PWM inverters may therefore be required to regulate the local bus voltage. In such cases the node has to be treated as a *PV* bus.

Slack, Swing or Reference Bus

The two types of buses described earlier require that the Ps are specified at all the network nodes before the load flow calculations are initiated. This does not make sense in terms of the conservation of power principle because the transmission losses in the system are not known before the solution is arrived at! This conundrum is resolved by allowing one bus that has a generator connected to it to be specified in terms of the magnitude V and angle δ of its voltage. Such a bus is known as a *slack, swing or reference bus*. The voltage at this bus acts as the reference with respect to which all other bus voltages are expressed. At the end of the load flow the calculated P and Q at this bus take up all the slack associated with the losses in the transmission.

5.6.5 The Load Flow Calculations

In small networks, it is often possible to obtain valid and useful results by direct application of the mathematical analysis presented earlier. Also, larger networks can often be reduced to equivalent circuits that can be solved in the same way. However, load flow analysis of any system beyond a few nodes is carried out by a computer.

Each bus of a power network is characterized by a number of variables. For a network with predominantly reactive transmission line impedances all these variables are linked by the complex Equation (4.7a), replicated below:

$$S_B = -\frac{V_A V_B}{X_t}\sin\delta + j\left[\frac{V_B}{X_t}(V_B - V_A\cos\delta)\right] = P_B + jQ_B \qquad (4.7a)$$

This equation describes the performance of a synchronous generator but it is equally applicable to a transmission line linking two buses. Note that there are four variables associated with each bus: the active and reactive power injected or extracted at the bus and the magnitude and angle of the bus voltage. The complex Equation (4.7a) can be split into two separate equations, one for the real part and another for the imaginary part. For a network with n nodes, there are $2n$ simultaneous equations to be solved; hence at each node two of the four variables have to be specified. All this fits neatly with the earlier arrangement of specifying buses in terms of P and Q, P and V or V and δ.

The solution of the $2n$ equations describing the network is not a trivial task! The basic Equation (4.7a) linking the node variables is nonlinear because it contains products of the variables and trigonometric functions. The load flow solution to such nonlinear simultaneous equations requires iterative techniques and is beyond the terms of reference of this book. In general, therefore, the solution of most practical networks requires a very structured approach to the mathematics and is carried out by computers. Indeed, it was one of the first useful applications of computers during the 1950s. There are many mathematical techniques for performing load flow analysis. They are all iterative, and most begin by representing the network as a matrix. They often use Newton–Raphson or Gauss–Seidel iterative methods or embellished versions of these. They should all converge to very similar answers and the choice between the methods is mostly a matter of speed and reliability of convergence. The different methods can perform better or worse with different types of networks. For more information consult power system simulation packages PSS/E and DIgSILENT. A recent trend is that load flow software is being integrated into geographical information systems (GIS) and this integrates with the asset management side of the electricity utility business.

The assumption of constant voltage at the PQ buses inherent in the formulation of the load flow described above is not valid at the distribution level. The distribution transformers (11 kV to 400 V) shown in Figure 5.11 are not controlled and representing them as constant P and Q will lead to some inaccuracy. Load flow software that is intended for modelling the lower voltage levels of distribution networks often includes the facility to define loads that do vary with respect to voltage. However, obtaining load data that includes this characteristic may be less straightforward.

5.6.6 Results

The basic results provided by a load flow analysis are the power flows (both active and reactive) and the voltages and currents (both magnitude and phase) for all the lines and nodes throughout the system. (Strictly speaking, some of this information is redundant.) From these basic results, an engineer can readily identify or calculate:

- any overloaded lines or transformers;
- any voltages that are under or over acceptable limits;
- power losses, both active and reactive, for individual lines or for the whole system.

This information influences the design and operation of the network. Load flow studies are routinely performed to check the state of the network as the demand varies through the day or whenever changes to the network or to generation are being considered.

5.6.7 Unbalanced Load Flow

Ideally, the voltages and currents in three-phase power systems are perfectly balanced, which leads to efficient operation and greatly simplifies any analysis. However, in low voltage networks particularly, the phases are not always balanced in practice and full analysis requires unbalanced load flow techniques. Unbalanced load flow usually employs the method of *symmetrical components*, which involves the description of the unbalanced network as the sum of three simpler networks. Many software packages can perform unbalanced load flow

Figure 5.13   A 132 kV circuit breaker. This large and expensive device serves the transformer shown earlier in Figure 5.2. (Courtesy of Murray Thomson)

analysis (perhaps sacrificing other features), but in practice the data they require is often unavailable.

5.7 Faults and Protection

Faults in power systems can be caused by a whole range of events. Lightning strikes, overhead lines breaking in storms and workmen digging through cables are common examples. Faults often result in the conductors being effectively shorted together or shorted to ground. In either case, very large currents can flow. The system must be designed to withstand these currents for short periods and to disconnect the faulted section quickly, so that the rest of the system continues to operate with a minimum of disturbance. Faults occur quite frequently (very frequently during storms) and must be effectively managed in order to achieve good reliability for the system as a whole. The detection and management of faults in power systems is known as *protection* and is a highly developed aspect of power systems engineering.

At low power levels, fuses are used and conveniently provide both fault detection and breaking of the circuit. At higher power levels, *protection relays* detect faults and trigger *circuit breakers*. Figure 5.13 shows a typical circuit breaker.

5.7.1 Short-circuit Fault Currents

The effective management of faults requires knowledge of what currents will flow under all likely fault conditions. First, this knowledge is required in order to design the fault detection mechanism, which has to differentiate clearly between normal load currents (including surges when equipment is first switched on) and fault currents. Second, the magnitude of the possible fault current must be known in order to select the device that has to break the current.

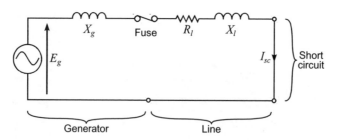

Figure 5.14 Single-phase equivalent circuit of a synchronous generator feeding a symmetrical short-circuit fault

Short-circuit faults can occur in a number of ways, including line-to-ground, line-to-line, etc., but, for simplicity, it is usual to start by considering the worst case of symmetrical three-phase faults. Fortunately this case is also the easiest to analyse.

5.7.2 Symmetrical Three-phase Fault Current

A symmetrical three-phase fault occurs when all three conductors of a three-phase line are shorted together. This amounts to the same thing as if they were all shorted to ground, but, because it is symmetrical (balanced), no current will flow in the ground and a simple single-phase analysis can be used.

In the simple case shown in Figure 5.14, E_g is the internal phase-to-neutral EMF of a synchronous generator, X_g is its internal per-phase reactance and R_l and X_l are parameters of the line. By inspection, the fault current I_{sc} is given by

$$I_{sc} = \frac{E_g}{|jX_g + R_l + jX_l|} \tag{5.9}$$

Obviously, if the fault occurred nearer to the generator, the total impedance would be less and the fault current would be greater. This equation gives I_{sc} as a scalar value, which is normal practice. For some purposes, however, the X-upon-R ratio of the impedance is also calculated.

5.7.3 Fault Currents in General

In a large power system, of course, there are many synchronous generators and many transmission lines interconnecting them. The *prospective fault current*, at any point on the network, will be the sum of contributions from all the generators. This prospective fault current will be different at different points of the network. Additionally, in practice, many faults are not symmetrical and a full analysis requires calculation of the unbalanced currents, usually employing symmetrical components.

The calculation of fault currents is mathematically similar to load flow and most power system simulation software packages will include *fault analysis* as standard.

5.7.4 Fault Level (Short-circuit Level) – Weak Grids

In systems operating at various voltages, it is useful to multiply the fault current I_{sc} by the nominal operating voltage (before the fault). On a per-phase basis this is $V_{ph}I_{sc}$, where V_{ph} is phase-to-neutral voltage. In a balanced three-phase system, $V_{ph} = V/\sqrt{3}$, where V is line-to-line voltage. Multiplying by 3 gives the three-phase fault level S_k:

$$S_k = \sqrt{3}VI_{sc} \tag{5.10}$$

where V is the nominal line-to-line voltage (before the fault) and I_{sc} is the symmetrical three-phase fault current, as introduced above.

The fault level is a slightly abstract quantity since the voltage and current being multiplied together are not occurring at the same time. It does *not* indicate the amount of power available. Its units are VA, kVA or MVA, not kW or MW. The usefulness of the concept is that it gives the rating of a circuit breaker capable of withstanding the full voltage when the system is normal and of interrupting the highest prospective fault current at the point at which the fault level is calculated. As a consequence the numerical value of the fault level and therefore the rating of circuit breakers is very large indeed. Table 5.2 gives typical values of fault levels at various system voltages.

The fault level at a particular point will vary as generators are brought on or taken off line elsewhere in the network. Also, it will vary if the network is reconfigured. Given that the entire protection system is designed around calculated fault levels, network operators must be aware of changes in fault levels and manage these appropriately.

The fault level is an important design parameter, not only for predicting currents under fault conditions but also for predicting performance under normal operating conditions. The Thévenin equivalent circuit in the following section illustrates this concept.

5.7.5 Thévenin Equivalent Circuit

Consider again the Icelandic national power system shown in Figure 5.12. It is often useful to be able to calculate performance at one particular node, without having to perform a full load flow or fault analysis of the entire network. This can be achieved by representing the network by its *Thévenin equivalent*, as explained in the Appendix.

The values in the Thévenin equivalent circuit shown in Figure 5.15 must be calculated for the particular node in question. The magnitude of Z_{th} can be found from the fault level at the node by

Table 5.2 Typical fault levels

Nominal system voltage (kV)	Fault level (MVA)
132	5 000–25 000
33	500–2 500
11	10–250

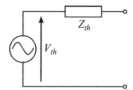

Figure 5.15   Thévenin equivalent circuit

$$|\mathbf{Z}_{th}| = \frac{V^2}{S_k} \qquad\qquad (5.11)$$

where V is the nominal line-to-line voltage. The angle of $\mathbf{Z}_{th}$ could also have been found from the fault level, if only $\mathbf{I}_{sc}$ and S_k had been calculated and stored as complex numbers. Instead of this, normal practice is to express the fault level as a scalar and to express the associated angle as an X/R ratio, which can be used in $\mathbf{Z}_{th} = R_{th} + jX_{th}$. The Thévenin source voltage can often be taken as the nominal voltage at the point of interest, being careful to use the phase-to-neutral or line-to-line value consistent with the calculation.

The fault level is an important design parameter, not only for predicting currents under fault conditions but also for predicting performance under normal operating conditions. It defines the *strength* of the network at a particular point. A *weak grid* is a network or part of a network where fault levels are low; i.e. the Thévenin or source impedance $\mathbf{Z}_{th}$ is high. A high $\mathbf{Z}_{th}$ implies that the node voltage would be sensitive to active and/or reactive power extraction or injection at the node.

The impact of a renewable energy generator on a network is very dependent on the fault level at the point of connection. Considering a proposed wind farm capacity (in MW) as a percentage of the fault level (in MVA) can provide a rough guide to acceptability. This topic is dealt in greater detail in Section 6.1.3.

5.8 Time Varying and Dynamic Simulations

One load flow calculation (after it has iterated and converged) provides the results for one snapshot of power system operating conditions. Many power system simulation packages allow the user to define load and generation profiles: time series data at daily, hourly or shorter intervals. The load flow is then run repeatedly to provide voltage profiles, loading profiles, etc., on the same time-base.

Reducing the time-step to milliseconds allows *dynamic simulations* to be performed, in which the dynamic characteristics of the generators and their associated control systems are included. This type of dynamic simulation is used for stability studies, particularly in the design of transmission systems. The behaviour of the network is modelled through repeated load flow calculations similar in principle to those discussed throughout this section. Load flow uses phasor representations of voltages and current and therefore implicitly assumes pure sinusoidal operation.

More sophisticated simulation packages offer *transient analysis*, in which the components are represented by differential equations. This allows nonsinusoidal operation to be modelled but is very much more computationally demanding. Transient simulation is used to model lighting and other fault conditions requiring very detailed analysis of systems with relatively few nodes.

Many power system simulation packages now include facilities to model renewable energy generators such as wind turbines and wind farms. A few years ago, the models tended to be little more than a 'negative load', which is adequate for basic load flow calculations but not much else. Today, the leading packages include much more sophisticated models of wind turbines, intended for use in fault analysis and dynamic simulations. However, the dynamic behaviour of wind turbines, particularly those using doubly-fed induction generators (DFIGs), is very dependent on the detail of the generator control software. The turbine manufacturers are reluctant to provide these details, which has hampered the development of software simulation models. Some simulation software houses claim to have obtained sufficient details, but validation of these models remains an issue.

5.9 Reliability Analysis

The ability of an electrical power system to satisfy the demand with a reasonable probability of continuity is a measure of reliability. Loss of electricity supply may result in significant economic and social impact to consumers. Maintaining a reliable supply is therefore of paramount importance to electric generators and distributors. The subject of power system reliability evaluation is one that has been studied extensively for traditional generating systems. The evaluation of reliability when distributed generators are also present complicates the issues and is a topic that is gaining importance. Distributed generators possess variability that may be beneficial; for example their statistical output is correlated to local demand or the converse, in which case they may contribute disbenefit under certain conditions. The general unpredictability that characterizes most RE sources requires that past reliability assessment techniques based on deterministic principles must be modified to account for the probabilistic nature of embedded generation. This topic is beyond the terms of reference of this book, but more information can be found in Reference [4].

References

[1] National Grid Company (NGC) website www.nationalgrid.com/uk/. In particular, the 'Seven Year Statement' is very informative and now includes a large section on 'Embedded and Renewable Generation'.

[2] Gross, C. A. *Power System Analysis*, John Wiley & Sons, Ltd, Chichester, 1986.

[3] Wallis, E. A. (ed.), *Modern Power Station Practice*, Volume L, *System Operation*, Pergamon Press, Oxford, 1991.

[4] Jenkins, N., Allan, R., Crossley, P., Kirchen, D. and Strbac, G. *Embedded Generation*, IEE Power and Engineering Series 31, Institution of Electrical Engineers, London, 2000.

6

Renewable Energy Generation in Power Systems

6.1 Distributed Generation

6.1.1 Introduction [1–9]

Figure 3.2 in Chapter 3 shows that the maximum demand of England and Wales, a typical developed European country, is of the order of 50 GW. Most of this power is supplied by gas, coal and nuclear fuelled, combined cycle and steam driven generator sets rated up to 500 MW. The entire peak load, therefore, could be met by a hundred or so sets. As most power stations have two, three, or four such generator sets only a few tens of power stations are required to supply the many millions of consumers that are connected to the system. This type of generation is known as *centralized generation* and is typical of large power systems worldwide. Centralized generation is normally connected to the transmission system.

The term *distributed generation* or *embedded generation* refers to smaller generators that are usually connected to the distribution network. Distributed generation includes:

- generators powered from renewable energy sources (except large scale hydro and the largest wind farms);
- combined heat and power (CHP) systems, also known a co-generation (co-gen);
- standby generators operating grid connected, particularly when centralized generation is inadequate or expensive.

The focus here is on renewables, but much of this chapter applies equally to all three types of distributed generation.

Smaller generators are not connected to the transmission system because of the cost of high voltage transformers and switchgear. Also, the transmission system is likely to be too far away since the geographical location of the generator is usually predetermined by the availability of the renewable energy resource. Such generators are *embedded* in the distribu-

Renewable Energy in Power Systems   Leon Freris and David Infield
© 2008 John Wiley & Sons, Ltd

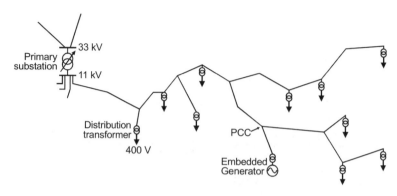

Figure 6.1   Example embedded generator

tion network. An example of an embedded generator in a distribution network is shown in Figure 6.1.

Distributed generation alters the power flows in distribution networks and breaks the traditional one-way power flow from the high to the low voltage layers of the power system. In some circumstances, a distributed generator can be accommodated into a distribution network without difficulty, but occasionally it can cause a variety of problems. This chapter examines these issues.

6.1.2 Point of Common Coupling (PCC)

A basic requirement in connecting any generator to a power system is that it must not adversely affect the quality of electricity supplied to other customers on the network. With this in mind, it is useful to identify again the *point of common coupling* (PCC).

Official definitions vary, but in simple terms the PCC is the point where the generator is connected to the public network as shown in Figure 6.1. In other words, the PCC is the point on the network, nearest to the generator, at which other customers are, or could be, connected. Thus, the exact location of the PCC can depend on who owns the line between the generator and the rest of the network. The importance of the PCC is that it is the point on the public network at which the generator will cause most disturbance.

6.1.3 Connection Voltage

The distribution system in the UK includes 400 V, 11 kV, 33 kV and 132 kV; other countries have similar voltage levels. As mentioned in Section 5.7.4, the fault level at the point of connection which is a measure of network strength, is an important design parameter, not only for predicting currents under fault conditions, but also for predicting performance under normal operating conditions and in particular, voltage rise.

The fault level at the PCC is very important when considering connecting a generator because it largely determines the effect that the generator will have on the network. A low fault level implies a high network source impedance, and a relatively large change in voltage

Table 6.1 Design rules. (Reproduced from Jenkins, N., Allan, R., Crossley, P., Kirchen, D. and Strbac, G., *Embedded Generation*, IEE Power and Engineering Series 31, Institution of Electrical Engineers, London, 2000. Reproduced with permission of IET)

Network location	Maximum capacity of embedded generator
Out on 400 V network	50 kVA
At 400 V busbars	200–250 kVA
Out on 11 kV or 11.5 kV network	2–3 MVA
At 11 kV or 11.5 kV busbars	8 MVA
On 15 kV or 20 kV network and at busbars	6.5–10 MVA
On 63 kV to 90 kV network	10–40 MVA

at the PCC caused by extraction or injection of active or reactive power. The impact of a renewable energy generator on the network is therefore very dependent on the fault level at the point of connection as well as on the size of the proposed generator.

The appropriate voltage at which to connect a distributed generator is largely dependent on its rated capacity. There are many other factors, as will be seen, and so a range of indicative figures are used as guidelines. Of course, whether a network is weak or not is entirely in relation to the size of the generator being considered. It is therefore common to express a proposed renewable energy source capacity (in MW) as a percentage of the fault level (in MVA) that can be labelled 'short circuit ratio'. This can provide a rough guide to acceptability. Typical figures for wind farms range from 2 to 24%.

Table 6.1 gives rough figures for the maximum capacity of a generator that can be connected at a particular voltage level. These rules should ensure that the influence of the generator on the voltage at the point of connection is acceptable. Connecting at a higher voltage is usually more expensive because of the increased costs of transformers and switchgear and most likely because of the longer line required to make connection with the existing network. Connecting at too low a voltage may not be allowed if the generator were to result in an excessive effect on the local network. This can lead to a situation where the developer of the renewable energy system wishes to connect at one voltage level for economic reasons, while the network operator suggests connection at the next level up.

6.2 Voltage Effects

6.2.1 Steady State Voltage Rise

The connection of a distributed generator usually has the effect of raising the voltage at the PCC and this can lead to overvoltages for nearby customers. The need to limit this voltage rise, rather than exceeding the thermal capacity of the line, often determines the limiting size of generator that may be connected to a particular location.

An initial estimate of the voltage rise caused by connection of a generator can be obtained from analysis of the system as represented in a simplified form by Figure 6.2. The network up to the point of common coupling can be represented as a Thévenin equivalent with the Thévenin impedance Z_{th} estimated from the fault level and X-upon-R ratio at the PCC.

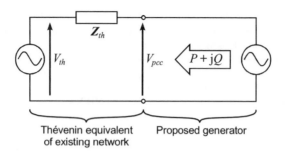

Figure 6.2   Equivalent circuit for estimating voltage rise

From Equation (5.11) the magnitude of Z_{th} is given by $|Z_{th}| = V^2/S_k$ where V is the nominal line-to-line voltage. The X-upon-R ratio is then used to find the resistance and reactance from $Z_{th} = R + jX$

The voltage rise ΔV can then be found using Equation (5.6): $\Delta V \approx (PR + QX)/V$, where P and Q are positive if the active and reactive powers are positive, i.e. have the directions shown in Figure 6.2. If an induction generator is connected directly to the network, Q has a negative value.

The allowable voltage rise is dependent on how the network is currently operated i.e. how close the existing voltages get to the allowable maximum. Typically, a rise of 1% would be a concern to the network operator. Voltage rise is often the main consideration for wind farms, which tend to be in rural areas, connected by long and relatively high impedance lines. Voltage rise often puts a limit on the amount of generation that can be connected in a particular rural area. This can occur long before there is any chance of the power flow actually reversing or of the thermal limits of the lines being reached.

In some situations, calculations (or load flow modelling) show that voltages will exceed acceptable limits for only a few hours in a typical year. In this case, it may be cost effective to constrain generation during those hours. The lost revenue may be small in comparison to the cost of installing a stronger line.

Voltage rise can be mitigated through the extraction of reactive power at the PCC. With induction generators (the norm for smaller wind turbines), this can sometimes be achieved by removing some or all power-factor correction capacitors. With synchronous generators (more common in hydro and biomass fuelled systems), the excitation can be adjusted. However, an embedded generator will normally be charged for any reactive power that it consumes except in low voltage networks (400/230 V) where reactive power is not normally metered. Thus, the normal initial design objective is to operate with a power factor near to unity.

6.2.2 Automatic Voltage Control – Tap Changers

The simple picture presented so far is complicated in practice by the existence of automatic voltage-control mechanisms in distribution networks. The most common mechanisms are automatic on-load tap changers, which are fitted to most transformers throughout the system, except the final distribution transformers (in the UK, the 11 kV to 400/230 V transformers as described in Section 5.5.1).

In the network shown in Figure 6.1, the primary substation is equipped with an automatic voltage controller, which uses a tap changer to adjust the turns-ratio of the transformer. Closed-loop control ensures that the voltage on the transformer secondary is kept close to 11 kV. This compensates for variations on the 33 kV side and the current-dependent voltage drops within the transformer itself.

Automatic tap changers affect the steady state ΔV voltage-rise analysis presented in the previous section. In particular, the Thévenin impedance Z_{th} must be adjusted, because its calculation was based on the fault level, which naturally included the impedance upstream of the voltage controller. With the automatic voltage controller active, the Thevenin voltage is the fixed voltage at the transformer secondary. To correct for this, the fault level and X-upon-R ratio at the fixed voltage node (transformer secondary) must be known so that the upstream impedance may be calculated and deducted appropriately.

To complicate matters further some automatic voltage controllers do not simply keep the voltage constant. For example, some provide line-drop compensation, which is a way of estimating and allowing for the voltage drop in the downstream line. Other controllers use a technique known as negative reactance compounding, which allows dissimilar transformers or transformers at separate substations to be operated in parallel. Unfortunately, controllers using these techniques can be affected by the changes in the substation power factor that can be caused by distributed generation. To prevent such changes in the power factor, some embedded generators are designed to operate at the same power-factor as the typical loads. Problems of this nature could be avoided if the distributed generators are of a type that could generate and control reactive power. In practice this option has not been fully exploited in the low voltage distribution networks, but with the ongoing development of more sophisticated and low cost power electronic interfaces it is likely to be of importance in the future.

6.2.3 Active and Reactive Power from Renewable Energy Generators

As mentioned earlier, the large generators of conventional power systems have their individual AVRs set to maintain the generator bus voltage virtually constant.

Generators fed from renewable energy sources are substantially smaller in rating and are, in general, connected to the distribution rather than the transmission network. For these reasons, conventional generator control schemes have not been considered appropriate for small embedded renewable energy supplied generators. For example, a fixed speed wind turbine driving an induction generator is expected to inject into the network whatever power is converted from the wind up to the rated wind speed, beyond which the power is limited by aerodynamic means at the rated output. It is also expected to absorb whatever reactive power the induction generator requires from the network, minus any locally generated Q from power factor correction capacitors. The situation is similar with small hydro systems interfaced to the grid through induction generators.

Energy from photovoltaic, wave and tidal schemes and MW size wind turbines will invariably be fed to the grid through a PWM power electronic converter. This provides the facility of reactive power injection/extraction at the point of connection.

To summarize, for relatively small embedded renewable energy generators connected to strong networks the P injection depends solely on the renewable energy source (wind, sun, water) level at the time and the Q injection either on the natural generator Q/P characteristics or on the control characteristics of the power electronic converter. In the latter case the converter could be regulated to inject active power at unity power factor, thus avoiding any Q exchanges with the network, and charges for reactive power. It is also possible that a mutually beneficial formula could be agreed between the owner of the renewable generator and the utility so that the Q generated/consumed by the converter is adjusted to suit the local network.

Generators that are very small individually have negligible influence on system frequency. However, when large numbers of such generators are connected their aggregate impact can be significant. In fact, as penetration levels of wind power increase, as for example in Denmark where the annual level exceeds 20%, and much higher levels will occur at times, utilities will require that renewable generators are designed to contribute positively to power system frequency and voltage stability during contingencies, but more of this later. In a similar manner, it might be expected that in future scenarios with substantial PV generation, that the PV converters might be required to be controlled in relation to system frequency.

6.2.4 Example Load Flow

As discussed already, renewable energy generation can affect both line loadings and voltages throughout the system. Load flow is a technique that allows the flows of real and reactive current throughout the network to be calculated, based on the location of the loads and sources and the line impedances.

The network of Figure 6.1 is used here to illustrate the way in which load flow analysis is applied at a distribution level to assess the effect of connecting a renewable energy generator at a node. This network is redrawn in Figure 6.3 to include node numbering and a proposed embedded renewable energy generator at node 10. It is necessary to know whether such an embedded generator is likely to affect adversely the network voltage profile.

At the outset a fault analysis is undertaken to provide the short-circuit levels at all nodes and so indicate the acceptable rating of the proposed renewable energy generator. To embark on a load flow it is necessary to specify the parameters of the lines, transformers and the known node variables. Table 6.2 gives the line data in terms of line position, length and line type. Table 6.3 provides additional information for each line type in terms of impedance and current rating.

Section 5.6.5 described the analysis required to determine network fault levels. For the network of Figure 6.3 the results are shown in Figure 6.4. As expected, the fault level is highest close to the secondary of the distribution transformer (node 1) and due to the cumulative impedance of the transmission lines declines moving away from this point. At node 10, where the renewable energy generator is to be connected, the fault level is 42 MVA. A wind turbine rated at 500 kW resulting in a short-circuit ratio of $0.5/42 = 12\%$ is within the acceptable range.

In this example it is assumed that the embedded generator is a wind turbine consisting of a directly connected induction generator operating at a power factor of 0.9 at full output. At

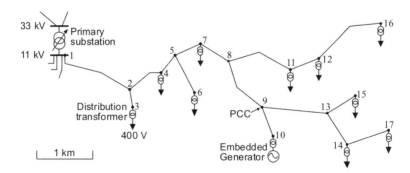

Figure 6.3   Example load flow

Table 6.2   Line data (input to load flow)

From node	To node	Length (m)	Line type
1	2	1297	100AAAC
2	3	304	50AAAC
2	4	626	100AAAC
4	5	391	100AAAC
5	6	738	185AL
5	7	492	100AAAC
7	8	583	100AAAC
8	11	1110	50AAAC
11	12	539	50AAAC
12	16	1253	185AL
8	9	1000	50AAAC
9	10	539	185AL
9	13	1154	50AAAC
13	15	583	50AAAC
13	14	652	50AAAC
14	17	791	95AL

AAAC: All aluminium alloy conductor

Table 6.3   Line type data (input to load flow)

Line type	R (Ω/km)	X (Ω/km)	Rating (A)
50AAAC overhead line	0.550	0.372	219
100AAAC overhead line	0.277	0.351	345
95AL underground cable	0.320	0.087	170
185AL underground cable	0.164	0.085	255

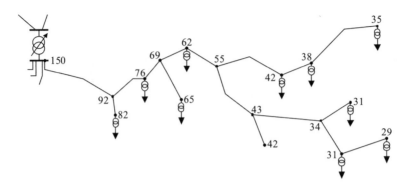

Figure 6.4   Example fault levels (in MVA) throughout a typical rural 11 kV feeder

Table 6.4   Node data (input to load flow)

Node number	P (kW)	Q (kVAR)
1		
2		
3	−238	−71
4	−159	−48
5		
6	−340	−102
7	−178	−53
8		
9		
10	500	−250
11	−458	−137
12	−221	−66
13		
14	−97	−29
15	−386	−116
16	−161	−48
17	−64	−19

a rated output this provides an injection of 500 kW and the absorption of 250 kVAR at node 10. In order to determine the voltage profile of the network under these conditions there is a need to assess whether this proposal is feasible. To proceed with the load flow analysis it is required that two variables are defined at each node. This distribution network consists solely of consumer *PQ* nodes and one embedded generator that can also be treated as a *PQ* node. The input data to the load flow are shown in Table 6.4.

The results of the load flow analysis are shown in Figure 6.5. The voltage profile at consumer nodes is easily within the acceptable limits of ±1%. At the PCC the rise of voltage due to the active power injection is in this case moderated by the extraction of reactive power from the renewable energy source. Of course, if capacitors are installed at the wind turbine

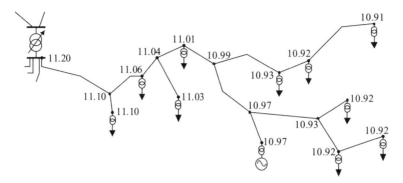

Figure 6.5   Load flow results: voltages (kV)

to correct the power factor towards unity, the voltage rise would have been significantly higher and perhaps unacceptable. A rerun of the load flow analysis would provide a wealth of information on system voltage sensitivity to a range of renewable energy source locations and operational conditions.

6.3 Thermal Limits

Power injected by a distributed generator will alter the local network power flows. Power flows in individual components may be increased, decreased or even reversed. Power flow (both active and reactive) requires current flows and this causes heating in lines, cables and transformers. Each of these components has a thermal limit, which effectively determines the maximum current it can carry, and these thermal limits can set an upper limit on the size of the renewable energy generator that can be accommodated. Load flow analysis allows all the component currents to be calculated.

6.3.1 Overhead Lines and Cables

It may appear surprising that a great many lines and cables are normally operated at well below their thermal limits, particularly in lower voltage networks (up to say 20 kV). There are two reasons for this. First, in order to keep customer voltages within range, voltage drops in conductors must be limited, which often requires use of larger conductors than those dictated by thermal limits. Second, energy losses are significantly less in larger conductors and a life-cycle cost analysis usually favours such oversizing. As a consequence thermal limits of lines and cables are rarely reported as a limiting factor in the installation of embedded generators in 11 kV or similar networks. Voltage and increased fault level considerations arising from the connection of additional generation, discussed later, usually come into play first.

At higher voltage levels, i.e. 33 and 132 kV, voltage control is less demanding and thermal limits of lines and cables may well be the limiting factor. These limits are well understood and rarely the subject of dispute, but there are a couple of aspects worth noting. First, care is needed regarding the demand that is assumed to be concurrent with the generation in cal-culations. Ideally this should represent the worst case scenario and does not necessarily occur when the maximum expected input from the renewable energy source(s) is concurrent with the maximum demand. Second, overhead lines are naturally cooled by wind, which increases their capacity during periods of high wind. Consequently lines installed primarily to carry wind power can be undersized relative to conventional design guides.

6.3.2 Transformers

Determining the thermal limits for transformers is more complicated. The rating of a trans-former is generally given in kVA or MVA, but there may be three different figures quoted, corresponding to natural (convective) cooling, fan cooling and cooling by pumped oil. Fur-thermore, transformers have long thermal time constants and thus can be overloaded for short periods of time without causing overheating or significant damage. Calculation of the damag-ing effect of such transient temperatures can be difficult.

Transformers are generally chosen to match the expected maximum demand and are nor-mally operated at a significant proportion of their thermal limit, partly because standing (no load) losses are significant and so the efficiency of a lightly loaded transformer is poor. Thus, in areas with very high penetrations of embedded generation, thermal limits of transformers can restrict further installation. In round figures, this only occurs when the rating of an embed-ded generator connected below a particular transformer, less the minimum load in the same area, exceeds the maximum load in that area.

Most transformers can accommodate reverse power flow, up to the normal forward rating, without a problem. However, there are some (not many) on-load tap-changers that have very limited reverse-power capability. Furthermore, some automatic voltage regulators (control relays) associated with on-load tap-changers can be affected by reverse power flow.

Thus, the thermal limits of existing transformers are not normally a limiting factor in the installation of distributed generators, unless the installed capacity of generators exceeds the maximum demand in the area served. Such high penetrations are not common, at least in the UK, although there is one example in Wales where two wind farms are connected below one substation. Power flow through this substation is frequently reversed and occasionally genera-tion has to be curtailed when the reverse power flow reaches the transformers' thermal limits.

6.4 Other Embedded Generation Issues

6.4.1 Flicker, Voltage Steps and Dips

Flicker

Incandescent light bulbs can cause annoyance by flickering due to rapid voltage variations. The human eye is particularly sensitive to flicker at around 8 Hz. Sensitive electronic equip-ment may also be affected. Such voltage variations are generally caused by rapid variations of active or reactive power flows within the network. The classic causes of unacceptable flicker are sawmills in rural areas and arc furnaces. Distributed generators can also contribute to flicker. A flicker meter measures voltage variations of this type and gives figures for short and long term severity.

Wind turbines can contribute to flicker because tower shadow and turbulent winds can cause rapid variations in active and reactive power output. Fixed speed pitch-regulated turbines are generally worse than stall-regulated ones at high wind speeds. This is because the pitch regulation action in response to, say, a gust is slow to take effect due to limited pitch actuator capability and the inertial mass of the blade, while the aerodynamic action of stalling a blade is very fast indeed. However modern multi-megawatt wind turbines are designed to operate at variable speeds up to the rated wind speed but also to allow marginal speed variations between rated and cut-out wind speeds. This results in smoothing of the power delivered, especially around rated power, and a substantial reduction in flicker.

The degree of flicker caused by a given generator is highly dependent on the network characteristics, in particular the fault level and X/R ratio at the point of common coupling. Thus, a generator that is perfectly acceptable at one point on a network may be unacceptable elsewhere. In general, this dependence is the same as that for a steady state voltage rise; thus, a wind farm that causes a significant voltage rise may also cause significant flicker. Flicker is more of an issue for single turbines or small clusters on low voltage distribution networks. As the power fluctuations from individual wind turbines are not correlated, the effect of flicker is less pronounced for large wind farms.

Steps and Dips

Switching a generator, either on or off, can cause a step change in the voltage at the PCC due to an associated abrupt change in active or reactive power flow. If the power changes associated with switching can be made more gentle, then such voltage steps can be minimized, at least during planned switching operations.

Direct on-line starting of an induction generator can cause a dip in the voltage, due to large starting currents, sometimes called inrush currents. For large induction generators, it is normal to use a *soft-starter* to limit these starting currents. In wind farms the practice is to connect turbines one at a time. Both steps and dips, if repetitive or cyclic, may contribute to flicker. With the increasing use of power electronic interfaces between the renewable energy generator and the grid, steps and dips during switching are virtually eliminated.

6.4.2 Harmonics/Distortion

Both loads and generators can cause distortion of network voltages, and such distortion is usually expressed in terms of harmonics. A common source of harmonics is the widespread use of cheap power supplies found in computers, televisions and other appliances. The voltage waveform in a modern office building or in a residential distribution network can be badly distorted by harmonics created by such loads, long before any distributed generators are even contemplated.

The distortion of the voltage results from nonsinusoidal currents flowing through the effective network impedance, which is related to the Thévenin impedance introduced earlier. The Thévenin impedance as calculated in Section 5.7.5 cannot be used directly to analyse harmonics because it is only accurate at 50 Hz. Experiment and analysis can provide a curve relating the system Thévenin complex impedance as a function of frequency. Such curves can be used in conjunction with the known levels of generated harmonics from say, a converter, to predict the likely distortion of the PCC voltage by a renewable energy embedded generator. Voltage distortion will usually be worse if the PCC node fault level is low.

In theory, a poorly designed directly connected synchronous or induction generator could cause significant harmonics, but modern machines manufactured by reputable companies are generally free of such deficiencies. Harmonics from distributed generators usually arise from their power electronics. Soft-starters for wind turbines are usually thyristor based and may cause significant harmonics, but only for the few seconds during start-up. This is unlikely to be a problem overall. Systems where the power electronics are used all the time are more of a concern. Some early wind turbines and PV systems used thyristor based converters and did cause significant harmonics.

Modern pulse-width-modulated (PWM) converters as applied to grid-connect PV and with variable speed wind turbines (using MOSFETs and IGBTs) are very much better in this respect but still have to be carefully designed and checked for suitability in each installation. In particular, there is a trade off because a high switching frequency will reduce harmonics but will increase switching losses and therefore reduce the converter efficiency.

A particular Spanish region with a very high penetration of wind power was found to have excessive harmonic distortion due to PWM converters with a rather low switching frequency. There is concern that very high penetrations of PV could have a similar effect, though this should be considered in the context of the rapid increase of power electronics in consumer appliances. National and International Standards, for example EN 61000-3-2: A, specify strict limits for the harmonic distortion allowed from small loads and generators, and network power quality in this regard is not regarded as a particular concern.

6.4.3 Phase Voltage Imbalance

Ideally, single-phase loads are evenly distributed between the three phases. In practice, any differences will cause a phase voltage imbalance on the network.

A directly connected three-phase generator will normally serve to reduce any existing phase voltage imbalance. However, in achieving this, the generator itself will heat up due to abnormal circulating currents within the windings. To avoid overheating, wind turbines usually include *imbalance protection*. In some rural networks, this imbalance protection can operate frequently and cause significant loss of generation.

In small distributed applications of RE, e.g. domestic wind turbines or photovoltaic, single-phase generators could increase phase voltage imbalance. However, if such applications were to be adopted in large numbers, statistically they would be evenly distributed on average among the three phases just as loads are and no problems should occur.

6.4.4 Power Quality

Voltage flicker, steps, dips, harmonics and phase imbalance may all degrade what is loosely described as *power quality*. Such undesirable occurrences can be caused just as much by loads as by generators. In general, it is the responsibility of the operator of the load or generator not to unduly affect the power quality for other users in the area. However, the impact of any load or generator is dependent not only on its own characteristics but also on the strength of the network to which it is connected.

Network operators are paying increasing attention to power quality. This is not specifically in relation to distributed generation but mainly because of regulatory obligations set up to protect increasingly discerning consumers from the generation of harmonics caused by the plethora of electronic devices on the network.

6.4.5 Network Reinforcement

In some situations, the voltage effects discussed above could be reduced by reinforcing the network and thereby increasing the fault level at the point of common coupling. For example, it is sometimes possible to upgrade the conductors of an existing overhead line, which is known as reconductoring. Alternatively, if permissions can be obtained, it is usually more cost effective simply to build a new line.

When faced with an application to connect an embedded generator, it is common for the local electricity company to claim that the nearest point for connection to the existing network is too weak and to require that a new line be installed in order to reach a stronger part of the network. The cost of installing such a line can be very significant in comparison to the size and cost of the generator itself and will normally fall to the owner of the generator, although rules for such network reinforcement vary from country to country.

6.4.6 Network Losses

Enthusiastic proponents of renewable energy sometimes claim that, because distributed generation is nearer to the loads, it will reduce network losses. Unfortunately, this is not always true.

High voltage transmission networks are generally very efficient; thus reducing power flows in transmission networks does not bring very large reductions in losses. In distribution networks, losses are proportionally far greater and so, for losses to be reduced to any significant extent, the distributed generation has to be located very close to matching loads. Furthermore, its availability profile in time should reasonably match that of the local demand.

6.4.7 Fault Level Increase

As noted previously, fault level can be used as a measure of grid strength and can be compared with proposed generator capacity to give an indication of likely effects on line voltage. However, as shown in Figure 6.6, the fault level can itself be increased by the connection of embedded generators, particularly where synchronous machines are involved. This does not invalidate the previous comments on voltage effects, but it does represent another factor to be considered in the connection of a generator.

Switchgear is sized to suit the network fault level. Existing switchgear therefore represents a limit to the amount of embedded generation that can be accommodated in a given area. It is not just the switchgear on the site of the embedded generator that is important as a synchronous generator will increase the fault level throughout the local network and so the rating of all local switchgear must be considered.

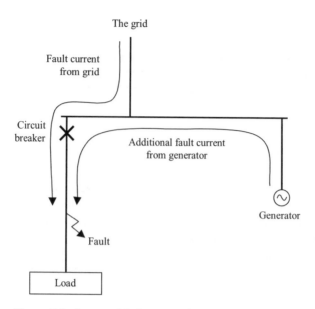

Figure 6.6   Increased fault current through a circuit breaker

In some cases, a proposed generator will make a relatively small contribution to the fault level but this could be enough to increase the existing fault level above the maximum allowable for a particular switchgear component. This component may be serving many other customers and be large relative to the generator. Upgrading this component could easily cost far in excess of the generator itself. A fault level increase is often the limiting factor for installation of embedded generation in urban areas where the lines are short and impedances low.

The good news is that the large majority of distributed renewable energy sources are not interfaced to the network through a directly connected synchronous generator but through a power electronic converter of the PWM type. Such converters consist of very fast switching elements and generally contribute negligible overcurrents during faults. In most cases there will be no need to upgrade the local circuit breakers.

Conversely, care must be taken to ensure that the inability of power electronic converters to provide fault current does not conflict with a protection system design that might rely for example on high currents to blow fuses.

6.5 Islanding

6.5.1 Introduction

A generator is said to be *islanded* if it continues to supply a local load after being disconnected from the main network. On a larger scale, an island can include many generators and many loads.

If islanding is to be permitted, then the system must be designed to be capable of operating both grid-connected and standalone. Generally speaking, the standalone mode is the more difficult to engineer since it must include generation/load matching in order to ensure that voltage and frequency are kept between acceptable limits. Generators that are designed to operate both grid-connected and standalone are often called *standby* generators and are installed primarily to provide security of supply to particularly important loads, such as hospitals and computer centres. Standby generators are usually powered from diesel or gas.

Systems employing renewables are rarely designed to operate both grid-connected and standalone. The extra cost of properly engineering a standalone mode is unlikely to be justified unless grid power cuts are very frequent.

If a generator is not designed to operate in a standalone mode, it must be capable of shutting down immediately when the grid connection is lost, otherwise the system may become dangerous for the following reasons:

- The island may not be properly earthed and may present an electric shock hazard to personnel.
- A system that appears to be disconnected from the mains but is in fact still live also presents a shock hazard.
- The voltage and/or frequency may deviate from the normal range and this may damage other equipment connected within the island.
- The island's frequency will quickly deviate from that of the main network and if the two networks are reconnected under these circumstances, equipment (particularly synchronous machines) will probably be damaged.
- The fault level within the island will be reduced and, if a fault were to develop, the resulting current might be insufficient to operate protective devices such as circuit breakers and fuses decisively.

6.5.2 Loss-of-mains Protection for Rotating Machines

Satisfactory *loss-of-mains* detection is surprisingly difficult to achieve with rotating machines. The detection must be reliable, fast and stable and not prone to wrong tripping. The requirement that it must be fast stems from the widespread use of *auto-reclose circuit breakers* in distribution systems and from the fact that these are not normally fitted with synchronization checking. Auto-reclosers are designed to restore a supply very quickly after a transient fault. Most faults in distribution systems (upwards of 80%) are transient and thus reclosers play a very important role in reducing customer-minutes lost. Reclose times in distribution systems can be as short as 1 second, which is good for consumers but also implies that loss-of-mains protection must be designed to operate well within 1 second.

For small induction generators, under/overvoltage and under/overfrequency protection is considered adequate protection against loss-of-mains (in the UK). For synchronous generators and large induction generators the preferred loss-of-mains detection techniques are:

- rate of change of frequency (ROCOF) and
- vector shift.

Both techniques rely on there being a significant change in power flow within the island when the mains is lost. This will naturally cause (a) the generator to accelerate or decelerate (detected by ROCOF) and (b) a change in the generator power factor (detected by vector shift). Unfortunately, there is no guarantee that there will be a significant change in power flow and so neither technique proves to be perfect in practice.

6.5.3 Loss-of-mains Protection for Inverters

An inverter that is intended for grid connection will not normally be designed to island. It will be designed to convert all the power available at a given time from the generating source (PV for example). It will not be designed to supply a constant AC voltage to a variable load. Islanding is only possible if the load demand exactly matches the source supply, a very unlikely scenario. Nonetheless, the designs of such inverters usually include explicit loss-of-mains protection, indeed connection standards such at the UK's Engineering Recommendation G83/1 [2] can require this.

6.6 Fault Ride-through

Prior to about 2001, all wind turbines were required to shut down immediately on the occurrence of any disturbance on the grid. This was primarily to reduce any risk of islanding. It was a sensible requirement in systems with very low wind power penetrations. Wind turbines were fitted with loss-of-mains protection relays triggered by any of undervoltage, overvoltage, underfrequency, overfrequency, rate of change of frequency (ROCOF) and vector shift incidents.

If such a protection relay is set to trip reliably on a genuine loss-of-mains, then it will also be prone to nuisance tripping when there is merely a transient disturbance on the mains. This raises the possibility of *block tripping* where a widespread transient disturbance across the network could cause a large number of loss-of-mains relays to trip simultaneously. Block tripping is highly undesirable in the context of controlling the voltage and frequency of the overall power system. The concern is that, following a disturbance on the grid, such as the tripping of a large generator, a substantial block of wind power would also trip off, which would aggravate the shortage of generation, making matters considerably worse for frequency control. Because of this, new regulations are now being drafted and introduced, particularly for very large wind farms that are to be connected to the transmission network. These regulations will require that wind turbines do not trip unnecessarily, but rather that they will ride through transient disturbances and to stay connected and resume generation as soon at the fault is cleared.

The first requirement would be for wind turbines not to be tripped off the grid during local faults that result in PCC voltage reductions, as is the practice at present. Wind turbines fitted with controllers capable of such behaviour have been tested and shown to be capable of riding through a fault and re-establishing full power immediately after the end of the fault. It is expected that increasingly sophisticated control functions are likely to be demanded from wind and other renewable energy generators as penetrations increase. Even with small but numerous embedded renewable generators in the future, it may eventually become important to coordinate their response both to steady state and transient network conditions. At present no such arrangements are in place.

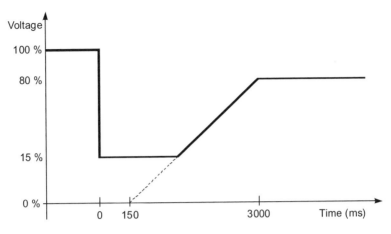

Figure 6.7   Fault ride-through voltage profile

In 2001, the German transmission system operator, E.ON Netz, issued new requirements including fault ride-through. Since then, most of the other major European transmission system operators have issued similar requirements, within their respective *grid codes*. The details vary between countries but the general requirement is as follows.

During a short circuit on the network near to the wind farm, the voltage at the connection point of the wind farm is assumed to follow the profile illustrated in Figure 6.7. The wind turbines must remain connected to the network; they should generate reactive power during the fault and should resume generation of active power as soon as the fault is cleared. Meeting this requirement was, and still is, a major challenge for manufacturers of wind turbines, especially with certain generator configurations.

Wind turbines with full converters are readily redesigned to meet the fault ride-through requirement. Turbines with directly connected induction generators are inherently unsuited. The drop in terminal voltage causes the flux to collapse, the electromechanical torque almost disappears and the wind turbine accelerates rapidly. To prevent overspeed the brake has to be applied and the turbine cannot immediately resume generation when the fault is cleared. To address this (and for other reasons), Siemens (formerly Bonus) added full converters to some of their models.

Doubly fed induction generator (DFIG) manufacturers also faced a considerable challenge. In order to protect the IGBT converters from excessive current during a fault, DFIGs are generally fitted with a crowbar, which is a set of thyristors that short out the rotor and cause the DFIG to behave like a squirrel-cage machine. All major manufacturers of DFIGs now claim to have developed their controllers such that they can ride through faults. In a competitive market, they do not publish the details, but it is understood that they generally employ an active crowbar, which provides a controlled short circuit to the rotor and allows the IGBT converters to continue operation.

The details of fault ride-through requirements differ from one country to the next, and there is concern that these details may be somewhat arbitrary. Meeting fault ride-through requirements adds costs to wind turbines and the industry is now concerned that these requirements should not be unnecessarily applied.

6.7 Generator and Converter Characteristics

Table 6.5 provides a summarized overview of the characteristics of synchronous and asynchronous generators and power electronic converters relevant to the issues discussed in this chapter. Further information on the topics of this chapter can be found in References [3–9].

Table 6.5 System-relevant characteristics of generators and converters

	Rotating machines		Power–electronic converters	
	Synchronous	Induction (asynchronous)	Line-commutated (old – thyristors)	Self-commutated (modern – PWM)
Operating speed (rev/min)	Exactly synchronous speed	Slightly above synchronous speed when generating, above and below synchronous in DFIG systems		
Reactive power	Either generate or consume; often controlled by an AVR, which may be set to give constant bus voltage or constant power factor.	Machine *itself* will consume; may have power factor correction capacitors fitted or may consume or generate if fitted with PWM converter	Consume	Either generate or consume; often designed to operate at unity power factor
Starting currents	Should be insignificant	Significant unless a soft-starter is used	None	None
Fault contribution	Significant	Usually insignificant	Insignificant	Insignificant
Synchronisation procedure required	Yes	No	No	No
Risk of islanding	High (if completely unprotected)	Low but slightly increased when power factor correction capacitors are fitted	None	Low (unless designed to)

References

[1] National Grid Company (NGC) website www.nationalgrid.com/uk/. In particular, the 'Seven Year Statement' is very informative and now includes a large section on 'Embedded and Renewable Generation'.

[2] ER G83/1 is available from the Energy Network Association; www.energynetworks.org.uk.

[3] McGrow, L. 'Beat the system', *IET Power Engineering*, April/May 2006, 28–30.

[4] Jenkins, N., Allan, R., Crossley, P., Kirchen, D. and Strbac, G. *Embedded Generation*, IEE Power and Engineering Series 31, Institution of Electrical Engineers, London, 2000.

[5] 'Decentralising power: an energy revolution for the 21st century', Greenpeace, 2006.

[6] Povlesen, A.F. 'Distributed power using PV', *Renewable Energy World*, March/April 2003, 63–73.

[7] Godoy Simoes, M. and Farret, F.A. *Renewable Energy Systems*, CRS Press, Boca Rabon, Florida, 2004.

[8] Masters, G.M. *Renewable and Efficient Electric Power Systems*, John Wiley & Sons, Ltd, Chichester, 2004.

[9] Fox, B., et al. *Wind Power Integration – Connection and System Operational Aspects*, IET Power and Energy Series 50, 2007.

7

Power System Economics and the Electricity Market

7.1 Introduction

This chapter reviews the economics of renewable power generation and the electricity market, which is increasingly the mechanism that determines what generation is used and when. At the outset a comparison is made between the capital and running costs of renewable and conventional forms of generation. Consideration is then given to how the economics change if external costs are taken into account along with the benefits/costs of small scale embedded renewable energy generation.

A description is then given of how renewable energy generation is financially accommodated within an electricity system in which power from a generation plant is dispatched to meet the varying electricity demand. The UK situation is reviewed to illustrate how renewable energy generation fits within a system that has evolved from being wholly state owned and controlled to a privatized market with open competition and trading.

Finally, the support mechanisms that are used around the world to encourage the growth of renewable energy generation are briefly reviewed.

7.2 The Costs of Electricity Generation

7.2.1 Capital and Running Costs of Renewable and Conventional Generation Plant

In order to assess the economics of renewable energy generation, it is first necessary to compare the costs of building and operating such plant in comparison to those of conventional generation. Installed costs and performance levels of renewable energy plant are broadly similar around the world, although no single figure can quantify the exact costs and performance of any technology. In the case of wind and solar energy, for example, the energy

Table 7.1   Indicative cost and performance data for renewable energy and conventional plant. (Courtesy of David Milborrow.)

Technology	Capital cost (€/kW)	O&M costs (€ cents/kWh)	Capacity factor (%)	Fuel cost (€ cents/kWh)
Onshore wind	1000–1500	0.9–1.5	20–50[a]	—
Offshore wind	1500–2000	1.5–3	30–40[a]	—
Biomass (energy crops)	1500–2700[b]	0.9–1.5	70–85[c]	0.5–1.5
Landfill gas	450–1300	1.3–2.7	70–90	—
Municipal waste	2300–6000	4.5–7.5[d]	70–85	[e]
Geothermal	1500–3000	0.75–2.3	75–85	
Photovoltaics	5000–7000	0.15–0.8	8–15	
Gas (CCGT)	450–700	0.3–0.8	85	2.3–3[f]
Coal	1000–1100	0.8–1.5	75–80	1.5–2.3
Nuclear	1700–2300	0.5–1.2	85–90	0.6–1.1

[a]Depends on wind speed.
[b]High cost plant is more efficient.
[c]Production falls after about 15 years.
[d]Depends on the mix of waste.
[e]Cost negative (~4.5–9 € cents/kWh), but steady supply needed.
[f]Gas prices are relatively volatile.

productivity (kWh per kW) of plant varies significantly with the resource. Solar installations near to the equator are generally more productive than those in, say, Sweden or New Zealand. Wind energy variations are more extreme; the windiest regions are New Zealand, the British Isles and Antarctica, while equatorial regions tend to have very low average wind speeds.

Table 7.1 summarizes the key parameters associated with renewable energy technologies and conventional plant so that a comparison can be made. The table also includes notes as to how these vary. *Fuel cost* is zero for most types of renewable energy but not for energy crops, where it is positive, nor for the waste-burning technologies, where it can be negative if landfill dumping costs are avoided. It must be emphasized that the costs quoted are intended to give an appreciation of a possible range of levels, but are not exact and do not apply to all applications.

Table 7.1 indicates the three components of electricity generation costs:

1. The capital cost includes the cost of the plant; land acquisition (unless a rent is paid, in which case this is a running cost); grid connection, although in some European states the utility has to bear the cost; and initial financing costs (as opposed to repayment costs). The way that capital costs translate into a component of generation costs is discussed in the next section.
2. Operating and maintenance (O&M) costs include insurance, rent and rates set by the local administrative authority, as well as the costs of labour and materials used for operations and maintenance.
3. Fuel costs.

7.2.2 *Total Generation Costs*

Capital costs are not a reliable guide to competitiveness. For example, low capital costs of combined cycle gas turbine plant does not result in low electricity generation costs unless the cost of fuel is also low. Similarly, although geothermal plant has a high capital cost it also has a high load factor and may therefore have similar generation costs to gas or wind.

National institutional factors have a major influence on the financing terms for borrowing money and on investors' expectations for the return on the equity they put into a project. In some places all the investment is provided by the state, or by nationalized industries, and so the equity contribution is zero. Irrespective of the sources of finance, however, a *cost of capital* can be derived that expresses the interest rate that can be applied to the total cost of the project to determine generation costs. Alternative terms for this parameter are *test discount rate* and *project interest rate*. Before privatization of the electricity industry in the UK the test discount rate was set by the Treasury and was 8% in 1990. In Denmark today, utilities almost invariably use *public sector test discount rates*, which are typically around 5–6%. In Britain and the United States, there are no fixed guides and project developers use criteria that tend to be strongly influenced by the financiers. The overall interest rate is, in practice, dependent on the relative proportions of debt and equity and the appropriate interest rates. A weighted average, appropriate for a project as a whole, tends to be around 11%, although higher and lower values are found.

The other parameter that strongly influences generation costs is the capital repayment period. The longer this is, the lower the annual payment to cover depreciation and interest and hence the lower the generation cost. In Denmark, this period often corresponds to the expected life of the project, where 20 to 25 years is common. In the private sector, depreciation periods also vary, but are generally in the region of 12 to 15 years.

The *annual charge rate* is the fraction of capital payable each year to cover repayments of the original investment, plus interest, and depends on both interest rate and repayment period. Typical values are shown in Table 7.2. Once the annual cost has been determined, the corresponding generating cost component is simply the annual cost divided by the annual energy generation.

Table 7.2 Typical annual charge rates. (Courtesy of David Milborrow)

Interest rate (%)	Repayment period (year)	Annual charge rate	Comments
5	20	0.0802	Used in Denmark
8	20	0.1019	Old 'nationalized industry' UK figures
10	15	0.1315	Possible present-day UK criteria
12	12	0.1614	Applied to investments if seen as more risky

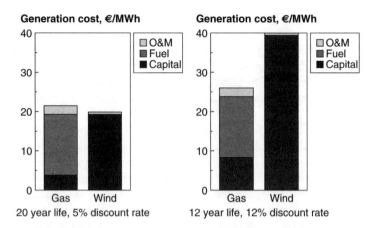

Figure 7.1   Generation costs as a function of financing terms. (Courtesy of David Milborrow)

Table 7.2 shows that the financing terms can have a very significant impact on the capital cost element of electricity generating costs. The most stringent criteria with the highest interest rate of 12% mean that annual charges will be double those associated with more relaxed criteria with the lowest interest rate shown at 5%. If the capital cost element of the total generation cost is small, e.g. as in the case of gas-fired generation, changes in the financing terms will have very little impact on generating costs. In the case of renewable energy technologies such as wind, wave, photovoltaics and tidal, where capital costs form the largest element of generation cost, the impact will be much greater.

Figure 7.1 illustrates this point, comparing generation costs for gas and wind, first with the financial criteria in the top row of Table 7.2 and second with the criteria in the bottom row. The results clearly illustrate the effect of moving from public to private sector financing. The prices move upwards, but the effect on a capital-intensive generation source, such as wind, is much more pronounced. The data on installed costs and performance levels were drawn from the Danish Energy Agency [1].

Only wind and gas are compared in Figure 7.1 whereas the IEA data [2] shown in Figure 7.2 cover a much wider range of renewable energy technologies. Again, the relative viability of the different sources is highly dependent on the financial criteria.

The IEA data do not include wave energy, currently the subject of increasing research activity in the UK and elsewhere, but commercial costs are inevitably very uncertain. The wave energy community suggests that a near-term generation cost target is around $80/MWh (€60/MWh).

7.3 Economic Optimization in Power Systems

7.3.1 Variety of Generators in a Power System

Power systems are in general fed by a variety of generators using a range of fuels. Each generator type has its own characteristics and is suited for a particular function within the system. Table 7.3 summarizes these characteristics and extends the features covered in Table 2.4 of Chapter 2.

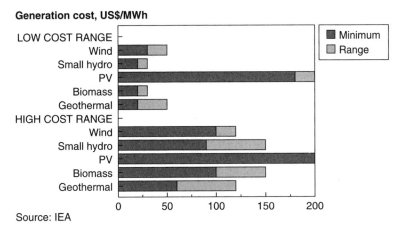

Figure 7.2 IEA estimate cost data for a range of RE technologies with high and low investment costs. (Courtesy of David Milborrow)

Table 7.3 Operational characteristics of traditional generation

Type of plant	Physical characteristics	Economic characteristics	Role within system
Nuclear	Inflexible; best operated at steady load	Very high capital costs; very low running costs	Steady base load; 75% capacity factor
Gas CCGT	Can be flexible, but with poor load following response	Low capital costs, low running costs; fuel supply contracts dictate running on high load	Steady base load; 80% capacity factor
Coal	Flexible with good load following response	Medium capital costs, medium–high running costs	Demand following, units partly loaded to provide reserve; capacity factor from 20 to 70%
Pumped-storage hydro	Extremely flexible	High capital costs, high marginal costs (see Section 7.3.3)	Rapid response used for 'peak shaving'
Hydro with reservoir	Flexible	High capital costs, low running costs	Base load and demand following if required
Open cycle gas turbine	Flexible	Very low capital costs, very high running costs	'Peak shaving'; capacity factor <<5%

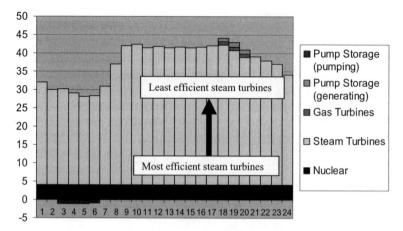

Figure 7.3 Typical generator scheduling to meet demand

Figure 7.3 shows that hourly demand throughout the day can be met by a variety of generating plant in a typical power system. Nuclear generation, being relatively inflexible, is allocated first. Pumped storage reservoirs are filled using low cost electricity during the nighttime period. More efficient gas and steam thermal generators are next in merit, and finally fast response gas turbines and hydro pump storage are used to meet demand during peak periods. In the next section, how the so-called 'optimum economic dispatch' is arrived at is discussed.

7.3.2 Optimum Economic Dispatch

The aim is to load generating sets in response to demand changes so as to minimize the cost of supply. This is known as *optimum economic dispatch* (OED) and over the years sophisticated OED methods have been developed to identify the minimum cost solution and at the same time satisfy numerous engineering and operational requirements and constraints. Here the basic ideas that provide guidance in the task of adjusting the load on individual conventional or RE generators will be investigated in order to achieve a minimum production cost on a short term basis.

A crude way of applying economic dispatch is to assume that the cost of operation of a plant is a linear function of the plant loading level, i.e.

$$C_A = K_A + K'_A P_A$$

where C_A is the cost in £/h for the Ath plant, K_A and K'_A are constants related to the capital and fuel plant cost components respectively and P_A in MW is the power delivered by the plant. To determine the optimum load sharing between two plants A and B when the total demand on the system is P_d the linear cost curve for unit A is plotted normally as in Figure 7.4, while the axis of the unit B curve is rotated by 180°. The origin O_A of unit A is placed at a distance from O_B equal to the total load $P_d = P_A + P_B$.

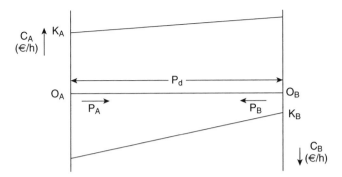

Figure 7.4   Demonstration of merit order loading

Any vertical line linking the curves will represent the cost of a given operating condition. A sharing of the demand by the two generators is sought such that the cost is minimum. From the diagram it is obvious that the minimum cost C_{min} occurs when the unit A takes as much of the load share as possible, i.e. it is loaded to its maximum level, with unit B brought in to cover the deficit.

The *incremental cost* of a plant is provided by the slope of the cost curve, $dC/dP = K'$. In the example of Figure 7.4, unit A might be a nuclear plant with very large capital costs (hence large K_A) and very low fuel costs (hence low incremental cost K'_A), while unit B could be a coal fired station with lower capital costs but higher incremental costs.

A plant loading methodology is now clear. If there are n plants available arrange the plants in order of increasing K', the so-called *merit order*. Starting from the top of the list, the plants are then loaded (each up to the limit of its capacity) before the next plant on the list is brought into action. This loading results in the scheduling shown in Figure 7.3.

To summarize, in the merit order approach, units with low operating costs run preferentially and therefore attract high load factors; they generate a disproportionately large share of electricity relative to their capacity. They are called *base load plants*, or high merit plants. Units with high operating costs only run during peak demand, generate a disproportionately small share of the total electricity and are known as *peaking plant*. Plants in between these two extremes are called *intermediate plants* or middle merit plants.

7.3.3 Equal Incremental Cost Dispatch

The merit order philosophy is adequate and convenient for rough scheduling but as a precise tool of economic dispatch it is inaccurate. The reason for this is that cost curves are not quite linear as assumed above; they are better described by a quadratic function as in the example of Figure 7.5. Such curves can be used to determine the overall efficiency of the plant at various output levels.

As previously, it can be assumed that only two plants A and B are supplying a demand P_d. Following the plotting method in Figure 7.4, the curve for unit A is plotted in Figure 7.6 conventionally while the axis of the unit B curve is rotated by 180°. Again the origin O_B of

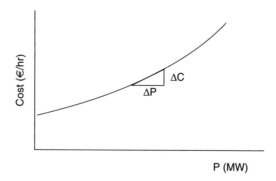

Figure 7.5   Typical input–output curve for a thermal generating unit

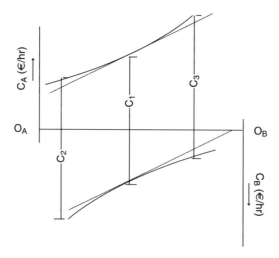

Figure 7.6   Demonstration of the equal incremental cost solution. (Based on figure in Neuenswander, J.R., *Modern Power Systems*, Intertext Books, 1973)

unit B is placed at a distance from O_A equal to the total demand $P_d = P_A + P_B$. Any vertical line intersecting the curves will therefore represent the cost of a given operating condition. As before a sharing of the demand by the two generators is sought such that the cost is minimum.

Let the vertical line C_1 represent the operating point at which the slopes of the two curves are equal. The length of C_1 represents the total cost per hour of operating both units. Any other operating points (such as the ones represented by lines C_2 or C_3) result in higher cost. Hence the most economic operating point is the one at which the slopes of the cost lines (i.e. the *incremental costs*) of the two units are equal. This result is valid no matter how many generators are feeding power to the system. The same conclusion can be proven using the

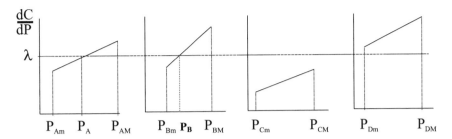

Figure 7.7 Economic distribution among four generators. (Based on figure in Neuenswander, J.R., *Modern Power Systems*, Intertext Books, 1973)

mathematical optimization technique of Lagrange multipliers. Note that another term for incremental cost is *marginal cost*.

7.3.4 OED with Several Units and Generation Limits

Steam driven generators have not only an upper limit of loading that corresponds to their nameplate rating but also a lower limit that is imposed by cavitation, which occurs in the turbine if the steam throughput falls below a certain critical level. This lower limit may be anything between 30 and 50% of the generator rating.

Differentiation of quadratic cost curves gives a linear function. Figure 7.7 shows incremental cost curves for a four-generator power system (with upper and lower limits for each generator) supplying a demand $P_d = P_A + P_B + P_C + P_D$. It can be seen that the whole range of incremental costs of generator C are below that of the other units; hence C should be loaded up to its full capacity P_{CM}. The new demand $P_d' = P_a - P_{CM}$ has to be distributed among the other three generators according to the equal incremental cost criterion.

Note that unit D, with a higher range of incremental costs, might be shut down (if units A and B are capable of providing the required reserve when partly loaded) or might also be required to provide reserve, in which case it will be loaded at its lower limit P_{Dm}. Assume that enough reserve can only be provided if unit D is loaded at its lower limit. The new demand then is

$$P_d'' = P_d' - P_{Dm}$$

which has to be optimally shared between units A and B.

Now choose a horizontal line (shown as λ in the figure). Suppose that the height of that line is such that it intercepts the incremental cost curves so that

$$P_A + P_B = P_d''$$

This is the graphical solution to the optimal economic dispatch problem in which the incremental costs of the generators operating between their limits are the same and equal to λ. The practical inadequacy of the merit order scheduling should now be obvious. The difference between the incremental costs of a plant at its minimum and maximum operating levels may exceed the difference between its average operating cost and the average operating costs of the plant on either side of it in the merit order list. The merit order technique is therefore an inferior tool for economic dispatch.

It should be appreciated that even if the total cost of generated energy by a power station, for example a nuclear power station, may be higher than that from a coal fired one, the OED requires that the nuclear station is fully loaded because of its very low incremental cost. The rationale is that the loading priority is independent of the original capital sum expended in the construction of a power station. As the station is present and available it should be utilized fully because of its low running costs. Whether the station should have been built in the first place is another matter. It may be concluded that as the energy supplied by RE sources (except biomass) is characterized almost entirely by a capital cost component, their incremental cost is negligible. Therefore the OED method would require them to provide at all times all the energy they are capable of delivering.

7.3.5 Costs on a Level Playing Field

Although generation costs are used to compare renewable energy and fossil generation, that process is not precise. A so-called 'level playing field' demands that allowances are made for various factors, some of which add and others subtract value to renewables. Essentially, the benefit of a new renewable source of energy for an electricity network may be specified as (value due to fuel savings) + (value due to reduction in external costs) ± (embedded generation benefits/disbenefits) + (value of capacity credit) − (costs associated with variability), where:

- The *fuel saving value* is simply the value of each kWh of fuel saved. In the UK in 2005, this amounted to about £15–20/MWh. The value varies on an hourly basis so this is a rough estimate of the annual average. The concept of capacity credit and the penalties due to variability were discussed at length in Chapter 3.
- *External costs* are costs attributable to an activity that are not borne by the party involved in that activity. All electricity generating technologies have associated external costs, and those of the fossil fuel sources of generation are due to the pollution that arises from their use and from the impacts of global warming due to their CO_2 emissions. Economists argue that these costs should be added to the generating costs, which is the thinking behind carbon taxes. However, these would add unacceptable increases to the price of electricity and so most governments give renewable energy sources a 'bonus' instead.
- *Embedded generation benefits/disbenefits* acknowledge that many renewable energy sources are small-scale and so connect into low voltage distribution networks. This means that losses in the electricity network may be reduced and, possibly, transmission and/or distribution network reinforcements delayed or deferred. The calculation of these benefits is a complex issue (see, for example, Reference [3]) and varies both regionally and locally. It is important to recognize, however, that, as mentioned in Chapter 6, concentrations of embedded generation can increase distribution losses in rural areas where demand is low and so may result in disbenefits.
- *Value of capacity credit and extra balancing costs* due to the RE source variability.

These topics were introduced in Chapter 4. All of these key components are discussed in the following sections.

7.4 External Costs

7.4.1 Introduction

Although there is general agreement as to the broad definition of external costs as costs attributable to an activity that are not borne by the party involved in that activity, there are widespread variations in defining the boundaries. For example, some have proposed that a substantial proportion of Western defence budgets should be regarded as an 'external cost' of oil. In light of the Gulf Wars, this may not be unreasonable.

External costs can be difficult to quantify but they can be highly significant. If the enormous costs of the clean-up operation after the Chernobyl nuclear disaster had been taken into account when the plant was constructed, it clearly would never have been built. External costs are set to play an increasing role in shaping future energy policy. Cleaning up the business of power supply is now a fundamental aim of many governments around the world. They are now recognizing the substantial costs to society of pollution, costs to which the electricity industry is a major contributor. The task facing energy policy makers is how best to go about the job of reducing pollution in electricity generation when in most countries external costs are not reflected in the market price of the end product. If this was done realistically, fossil fuel technology and nuclear prices would rise, making renewable energy sources more competitive.

Deregulation of the power markets can either aid or hamper the quest for proper recognition of external costs depending on the details. Crucial to the process is the willingness of governments to mandate that all energy options should compete on an equal footing. This spurs utilities to take the full costs of electricity supply into account. Such levelling of markets will also force the hidden subsidies to conventional technologies into the open; this *unbundling* is a primary aim of privatization, or deregulation. What is needed are fair and workable procedures to achieve what is referred to as the *internalization* of external costs.

7.4.2 Types of External Cost

Before looking more closely at the procedures available, a brief analysis of the make-up of external costs will help explain why they engender controversy. Table 7.4 gives a list of the local and general external costs excluding global warming.

External costs associated with energy supply/use are complicated. For simplification they can be divided into three broad categories:

- hidden costs borne by governments;
- costs of the damage caused to health and the environment by emissions other than CO_2;
- the costs of global warming attributable to CO_2 emissions.

The first category includes the cost of regulatory bodies and pollution inspectorates (generally small) and the cost of energy industry subsidies and research and development programmes.

Table 7.4   Sources of external cost due to electricity production, excluding global warming

Local	General
Smut deposition	Acid rain damage
Obscuration of the sun by plumes	to trees and crops
Noise due to plant, coal handling, etc.	to buildings
Noise due to fuel deliveries	to human health
Discharges into watercourses	to fisheries
Plant accidents – human cost	to animals
Smells	Oil spillages – clean-up costs
Dust and fumes	Ash disposal accidents
Heavy metal depositions	Heavy metal depositions
Upkeep of emergency evacuation measures (nuclear)	Leakage from radioactive waste

These can be very significant. In the ground-breaking analysis of external costs published by the European Commission in 1998 [4], it was calculated that support to the German coal industry added DEM 0.002/kWh to the price of electricity. A cost of DEM 0.0235/kWh was also assigned to nuclear research and development, compared to around DEM 0.004/kWh for wind power.

The second category covers costs due to emissions that cause damage to the environment and/or people. These make up about 10% of the external cost of power generation and include a wide variety of effects, including damage from acid rain and health damage from oxides of sulfure and nitrogen emitted from coal fired power stations. In a European study, ExternE [5], the costs of damage to health were estimated by calculating the loss of earnings and costs of hospitalization of people susceptible to respiratory diseases.

Other costs included in the damage and health category are power industry accidents, whether they occur in coalmines, on offshore oil or gas rigs, or in nuclear plant. The probability of a serious nuclear accident in Western Europe might be extremely low, but should a catastrophic failure occur the costs would be undeniably huge.

The third category is by far the largest: external costs due to greenhouse gas emissions which cause global warming with all its associated effects. This category accounts for some 40–100% of the hidden costs of the world's consumption of electricity. It is also the most contentious area of the external costs debate. The range of estimates for the possible economic implications of global warming is huge. Costs associated with climate changes, flooding, changes in agricultural patterns and other effects all need to be taken into account.

7.4.3 The Kyoto Agreements

The most high profile global agreements to curb man-made emissions of greenhouse gases, widely believed to be contributing to global warming, stem from the 1997 Kyoto Conference. The agreements emerging from this conference are obligatory for ratifying countries through the United Nations Framework Convention on Climate Change (UNFCCC), with associated penalties for nonconformance. These agreements form a basis for reducing the emissions of a 'basket' of greenhouse gases from the industrialized countries (referred to as 'Annex 1' or

'Annex B' countries) by 5.2% of the 1990 levels by a commitment period between 2008 and 2012. It is expected that this reduction will be achieved in each Annex 1 country through the 'domestic measures' of introduction of more efficient, and/or less carbon intensive, systems of power generation and industrial processing, with possibilities for the development and management of forests and agricultural soils ('carbon sinks') to absorb carbon dioxide from the atmosphere. These 'domestic measures' can be supported by the use of international 'flexible measures' such as emissions trading, joint implementation (JI) and the clean development mechanism (CDM). Countries signed up to Kyoto have introduced a range of policy measures aimed to ensure they meet their emissions targets. Some of these have the effect of internalizing at least some of the external costs associated with global warming.

7.4.4 Costing Pollution

The studies mentioned above along with several others have looked at overall damage potentials on health and the environment, assigning a cost penalty to each generating technology, depending on the fuel. This approach enables the difference in external costs between, for example, coal and wind, to be easily compared. Another approach calculates costs per pollutant, recognizing that different fuels generate different amounts of pollutant. Typically these penalties, as proposed by some states in the USA, and as shown in Figure 7.8, are around $10/tonne (€7/tonne) for CO_2, and up to $25\,000$/tonne (€9000/tonne) for SO_2. Since the quantities of each pollutant, per kWh, are well known for the differing fuels, these costs enable calculation of the external cost of each unit of electricity. Coal fired generation is highly polluting and produces about 1 kg of CO_2 per kWh, plus SO_2 and other pollutants. It therefore attracts the highest penalties, as shown in Figure 7.9. Despite the wide variety of approaches, the external cost assigned to coal fired generation generally adds about 1 USc/kWh. Gas, on the other hand, attracts a lower penalty. Table 7.5 summarizes data from the studies already cited, and others, such as References [6] and [7].

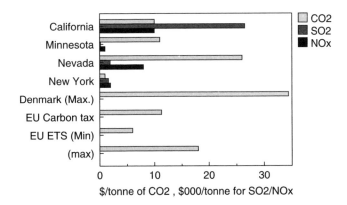

Figure 7.8 Taxes on emissions levied or proposed. (Courtesy of David Milborrow)

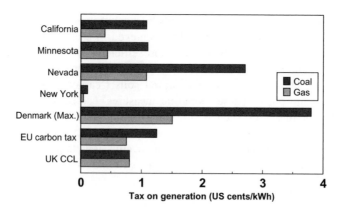

Figure 7.9   Taxes on generation levied or proposed. (Courtesy of David Milborrow)

Table 7.5   Estimates of external costs, in € cents/kW h. (Courtesy of David Milborrow)

Category	Coal	Oil	CCGT	Nuclear	Wind
Human health and accidents	0.7–4	0.7–4.8	0.1–0.2	0.030	0.040
Crops/forestry	0.07–1.5	1.600	0.080	Small	0.080
Buildings	0.15–5	0.2–5	0.05–0.18	Small	0.1–0.33 (1)
Disasters				0.11–2.5	
Total, damage	0.7–6	0.7–6	0.3–0.7	0.2–2.5	0.2–0.5
Global warming estimates	0.05–24	0.5–1.3	0.3–0.7	0.020	0.018
Indicative totals	1.7–40	3.7–18.7	0.83–1.86	0.36–5	0.4–1.0

7.4.5 Pricing Pollution

Costing pollution is not easy, but devising market mechanisms that take external costs into account is also problematic. The solution favoured by economists is *carbon taxes*. These would simply be added to the cost of electricity generated from particular sources so as to reflect the relevant external cost. The EU's proposals for carbon tax, for example, would have added about €18/MWh to the generation cost of coal and about €10/MWh to the generation cost of gas. Apart from the problems of administering such a tax, the impact on electricity prices was estimated to be too severe by most of the member states. The Netherlands and Denmark, among others, have nevertheless acted independently and introduced carbon taxes.

The EU's proposals for a carbon tax are unlikely to be implemented, but *carbon trading* commenced in 2005 [8]. The EU Emissions Trading System (ETS) sets limits on member states' carbon emissions from large industrial plants including fossil fuelled power stations. The intention of the ETS is that the limits for each country will be steadily reduced over a number of years to help deliver member states' obligations under the Kyoto Agreements. In practice, fossil fuelled power plants have an annual quota for carbon emissions and those who exceed their allowances must purchase 'credits' from generators or others who have not exceeded their allowances. In February 2005, allowances traded at just under €10/tonne of CO_2, which corresponds to a premium of €1/MWh on the price of coal fired generation, and about half that amount on gas fired generation.

Another approach to external costs – favoured by some American states – is to use *integrated resource planning* (IRP) to assess future generating plant needs. IRP is a process whereby the different options for meeting future electricity demand are considered and the least cost solution for delivering energy is selected within the constraints imposed by ensuring security of supply. The benefit of IRP is that, in complying with government mandates to consider all energy supply options, the regulatory body simply assumes that external costs are to be applied, without actually imposing any. The precise monetary value of pollution damage, or the calculation of cost penalties on polluting fuels, becomes less critical. This straightforward approach does away with many of the complicated mechanisms required to calculate the monetary value of pollution in electricity generation. It also helps to ensure that power plants are built in the right place and that the electricity network operates efficiently. The result is a competitive market operating under guidelines aimed at long term economic responsibility. However, this approach has not been without problems. Recent studies in California and elsewhere indicate that the net result of this approach is simply to shift the utility plant mixes in favour of gas, which is currently cheap and which has lower external costs than coal. Renewables profit, but the additional market share assigned to them so far remains relatively small. It is important to remember, however, that gas is unlikely to remain cheap indefinitely. In addition, government and popular pressure will accelerate the introduction of renewable energy technologies.

Popular consumer desire for more clean electricity is illustrated in the trend towards *green pricing* in the US, Australia and Europe. Given the choice, electricity consumers often agree to pay more for green electricity. In the Pacific Northwest of the US, a modest 4% increase in rates is making a utility 20% 'green' and the initial reaction of the consumers is 75% in favour. However, while green pricing is a clear signal to governments to promote renewables in the electricity market, it cannot be regarded as a substitute for an equitable energy policy. Furthermore, green pricing will not necessarily deliver the most efficient system overall. It is likely that a degree of IRP would need to be integrated into the market.

There are various other ways of reflecting external cost, directly or indirectly, and Table 7.6 gives a brief summary. It should be noted that these support mechanisms historically have tended to change on a frequent basis as electricity markets have evolved and the renewable energy industry has developed.

7.5 Effects of Embedded Generation

7.5.1 Value of Energy at Various Points of the Network

In the case of small scale renewables there is the potential for an *embedded generation benefit*. A simple example may be used to illustrate the point. If a householder were to install a photovoltaic panel on the roof of her/his house and export the surplus electricity to the local utility, they would resell it at between 11 and 14 € cents/kWh to neighbours. There would be no backflow through the local 33/11 kV transformer and, as a bonus, the locality might suffer fewer power cuts.

Table 7.6 Summary of support mechanisms for renewable energy. (Courtesy of David Milborrow)

Type	Country	Comments
Non-fossil fuel obligation	UK, IRL, FR	Generators made bids for renewable generation. If successful, they were given contacts for fixed terms
Feed-in tariffs	DE, ES, DK	Fixed payments made for renewable energy schemes. Tends to be expensive, but has been very successful in stimulating growth
Supply side obligations	UK, US states	Tends to be costly (UK), although this depends on precise rules
Production tax credit	USA	Fixed sum (1.8 USc/kWh or 1.4 € cents/kWh) added to generation from renewable plant. Works well, but depends on periodic renewals of mandate
Carbon trading	EU	Premium for renewables depends on quotas for fossil plant. Modest impact at present

This reasoning may be simplistic but it is nevertheless the principle behind *net metering*, which is used by several US utilities to remunerate renewable energy generators at 100% of the supply cost. A modified version of the approach has been used in Denmark and Germany where renewable generators were paid 80–90% of the domestic tariff.

Although studies in the US and elsewhere point to the fact that renewable energy may have a high value in certain locations [9], this point is not widely accepted within EU utilities, many of whom continue to assign renewable energy a value based only on its fuel saving capability. The cost, price and hence the value of electricity varies on a geographical basis. Above-average costs are incurred when energy has to be transmitted long distances from the generating plant. The need to supply remote locations increases the costs of the distribution network and, in addition, electrical losses are incurred in feeding the energy to the extremities of the system. In many instances renewable electricity generating technologies deliver energy closer to consumer demand than centralized generation. It substitutes for electricity that has accrued a higher selling price than the generation costs of large thermal plant, due to its passage through the network. This rising level of 'value' with reducing voltage is reflected in higher charges for customers connected at lower voltage levels. These concepts are illustrated in Figure 7.10, which includes typical selling prices for electricity at various voltage levels in the UK system.

Electricity utilities rarely ascribe exact costs to account for all geographical and temporal factors. Tariffs build in cost averaging so that blocks of consumers with similar requirements and locations pay similar rates.

7.5.2 A Cash-flow Analysis

As part of an EU-funded study [10], an analysis of the cash flows through the electricity network in England and Wales was carried out. Data were compared with information from

Electricity distribution and renewable energy

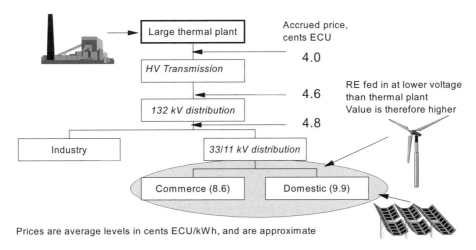

Figure 7.10   Principles behind embedded generation benefits.

tariffs, and with other case studies in Europe and the United States. The analysis enabled average levels of electricity price to be established at key points in the distribution chain:

- at the busbars of large thermal plant;
- at exit from the high voltage transmission system;
- at delivery to the consumer.

Conventional terminology delineates three stages of electricity supply:

- generation;
- high voltage transmission (often abbreviated to 'transmission');
- distribution.

For each of these stages in the supply chain, costs can be allocated under four headings:

- *fuel* (fuel cost purchases for generators and power losses in transmission and distribution);
- *operating*, reflecting all operating costs including overheads;
- *depreciation* of capital assets;
- *profit* on a historic cost basis.

Table 7.7   Segregation of costs in UK ESI, 1993/1994. Data in € cents/kWh. (Courtesy of David Milborrow)

	Fuel	Operating	Depreciation	Profit	Total
Generation	2.03	1.28	0.35	0.35	**4.01**
Control		0.14	0.01	0.03	**0.18**
Transmission	0.08	0.11	0.16	0.17	**0.52**
Distribution	0.31	0.80	0.42	0.58	**2.11**
Supply		0.26	0.01	0.13	**0.40**
Total	**2.42**	**2.59**	**0.95**	**1.26**	**7.22**

The cash-flow analysis was carried out for 1993/1994 and costs in nominal money values have remained similar since that time (which means they have fallen in real terms). The results of the cash-flow analysis are shown in summary form in Table 7.7. The entries for 'fuel' under transmission and distribution cover system losses. The price elements shown in Table 7.7 apply to the system as a whole. Regional variations occur for many reasons, primarily due to the proximity of a locality to sources of generation.

Although the constituent elements of electricity price shown in the table define its 'value' at any given point in the system in average terms, it does not follow that electricity which is injected at that same point will be credited with that same value by the generator or utility. A brief summary of current attitudes to value is as follows:

- *Fuel.* Most studies accept that fuel savings are a component of the value of renewable energy. With variable sources, the value declines as the amount of variable capacity increases.
- *Operating costs.* These include 'fixed' and 'variable' components. The variable costs of generation are part of the value of RE, but the variable costs of transmission and distribution are more contentious issues. Acceptance of the fixed component of these costs as a component of value depends on whether it is accepted that RE, particularly the variable sources, has a 'capacity credit', as discussed in Chapter 3.
- *Depreciation.* The crucial point is whether the introduction of RE on to a system enables plant savings to be made. Most studies have addressed generation plant savings, but more recent work addresses transmission and distribution savings.
- *Profit.* The question of profit as an element of value has not been explicitly addressed. In a deregulated system, where no single generator or distributor has exclusive rights, it would seem valid to allow its inclusion, at least partially, in value assessments.

7.5.3 Value of Embedded Generation – Regional and Local Issues

Transmission limitations may sometimes impose constraints [11] on the assimilation of new generation capacity. For example, in the case of wind in the UK, concentrations of wind in Scotland and the North of England, where wind speeds are higher than those further south, may cause an increase in the north–south power flows, which are already substantial [12].

Considerable reinforcement costs for the transmission networks would then be needed, depending on the timing and location of new renewable generation. Studies carried out by the network operators suggest that 6000 MW of new and renewable capacity in Scotland might trigger reinforcement costs of up to €2300 million [13]. The connection of similar amounts in England would trigger lower transmission reinforcement costs.

In practice, higher transmission connection charges in the North of England and Scotland may dampen the enthusiasm for wind projects. These are annual charges, paid by generators, that reflect the fixed costs of the transmission system associated with their plant. The indicative charge for plant in the Scottish Hydro zone is €30/kW [14], whereas plant in the southwest is paid €14/kW, reflecting the shortage of generation in that region. It should be noted that substantial reinforcement of the north to south transmission links might also take several years to implement.

Local issues are a complex topic (see, for example, Reference [3]) as they vary both regionally and locally. Concentrations of embedded generation can increase distribution losses in rural areas where demand is low, although modest amounts may reduce losses. A study of a ten-machine, 4 MW wind farm connected into an 11 kV system in Cornwall provided valuable information on local issues [15].

In view of the complexity of the issues involved and the fact that the impact is dependent on location, it is difficult to quote definitive values. Broadly speaking, however, the embedded benefits rarely exceed 2.3 € cents/kWh in the UK and are often much lower. Renewable generation may incur disbenefits where substantial quantities of such generation, for example in Scotland, may trigger the need for significant (and costly) transmission reinforcements. A UK report [16] examined the implications of installing up to 6000 MW of renewable generation in Scotland and concluded that reinforcement costs would be roughly €750 million per 2000 MW of generation.

7.5.4 Capacity Credit

An additional issue in the context of economic appraisals of intermittent renewable sources is the capacity credit of the source. This topic was discussed extensively in Chapter 3 and it is revisited here in the context of its economic implications. As mentioned previously, the capacity credit of any power plant may be defined as a measure of the ability of the plant to contribute to the peak demands of a power system. Numerous utility studies have concluded that wind can displace thermal plant.

The capacity credit of wind in northern Europe is roughly equal to the capacity factor in the winter quarter [17]. Results from ten European studies are compared in Figure 7.11, showing credits declining from 20–40% to 10–20% with low and 15% wind penetration respectively. It should be noted that the values of capacity credit depend on the capacity factor of the wind plant.

The UK National Grid Company has estimated that 8000 MW of wind might displace about 3000 MW of conventional plant and 25 000 MW of wind (20% penetration) would displace about 5000 MW of such plant. Figure 7.12 compares values of capacity credit normalized for annual capacity factor (as different values were used in the three studies) and shows a good measure of agreement. 'CEGB' refers to a study before privatization, 'NGC' a later study after privatization and 'SCAR' is referred to in Reference [18]. With modest contributions of

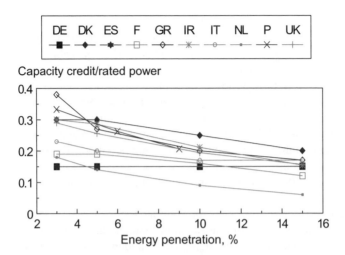

Figure 7.11 Comparison of results from ten utility studies of capacity credit, though assumptions differ. (Courtesy of David Milborrow)

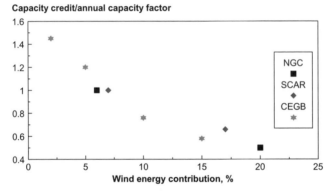

Figure 7.12 Normalized values of capacity credit from three studies of the NGC system. (Courtesy of David Milborrow)

wind energy the capacity credits are about 40% greater than the annual capacity factor and therefore, if the average capacity factor across the country was 35%, then 1000 MW of wind would displace 490 MW of thermal plant. At higher wind energy penetrations the capacity credit declines due to the requirement for more part-loaded generation plant to cover for uncertainties in wind power generation.

No evidence appears to exist to demonstrate that the conclusions set out above are inaccurate. Although there are periods of calm during the summer, the loss of load probability at such times is low. The nonavailability of wind during such periods therefore makes little difference to the year-round probability. It may be noted that the capacity credit of solar plant in the UK would be very low and that of tidal barrage plant would be lower than the average capacity factor.

Once the capacity credit has been determined, the value of capacity can then be estimated. The alternative new generation technology is mostly combined cycle gas turbines, costing around €700/kW, and the annual capital replacement cost of such plant is about €70/kW, although this depends on the annual cost of capital. At low wind energy penetrations the value associated with the capacity credit is around 0.8 € cents/kWh.

7.5.5 Summary

In the UK, the government has set a target of achieving 20% renewable energy in electricity supply by 2020. This provides a basis for estimating the total additional cost to the electricity consumer, taking into account all the above factors. An analysis completed in 2003 suggested that the extra cost would represent an additional 0.3 p/kWh (0.45 € cents/kWh) on consumer electricity bills [19]. As wind energy is expected to contribute the majority of the renewable energy target, the analysis assumed that all the renewable energy would come from wind and made projections about future trends in wind energy costs. Similar analyses have been made for the state of Pennsylvania and for the Republic of Ireland [20], with broadly similar conclusions. The Irish study suggested that electricity consumers might realize savings from about 2010 onwards when the additional costs associated with extra balancing were outweighed by the lower costs of wind compared to gas. In all these studies, however, it is essential to examine the assumptions carefully.

7.6 Support Mechanisms for Renewable Energy

7.6.1 Introduction

The previous section assessed the value of renewable energy generation when competing directly with other 'conventional' power generation sources in an open electricity market. Many countries have recognized that renewable energy would find it difficult to compete on level terms due to the issues of:

- Variability. As discussed, most renewable energy sources cannot easily be controlled and their output forecast with 100% accuracy.
- Cost. Many of the renewable energy technologies are not yet mature and thus cost more than conventional generation at present.
- External costs. The external costs of conventional generation (pollution, CO_2 emissions, etc.) are not yet properly accounted for and the flip-side of this is that renewable energy generation is not properly credited for avoiding these external costs.

The basic ways that governments have dealt with this in order to promote the uptake of renewable energy are:

- The introduction of a *feed-in law* where different renewable energy generation technologies are paid a fixed price for their output and suppliers/distributors are obliged to take that output.
- A *quota system* whereby suppliers/distributors are obliged to source a certain fraction of their energy from renewable energy sources. This is facilitated by the issuing of green certificates to renewable energy generators which can be traded with the energy or as a separate entity.
- A *carbon tax* on conventional energy generation, making renewable energy generation more attractive.
- *Carbon trading* (as described above).
- *Tax relief* on the investment in renewable energy generation plant.

This section will discuss the implementation of these schemes by looking at two European countries, namely the UK and Germany, who have implemented all of these measures in some way. An analysis is then made of the value of green certificates under a quota system followed by a brief discussion of the selling of green electricity as a 'product'.

7.6.2 Feed-in Law

Germany, like the UK, set a target in 1998 for 10% of electricity to be generated from renewable energy sources by 2010. Since 1991, a law has been in force which gave a legal obligation for utilities to take off all electricity generated by renewable energy generators at a fixed price depending on the technology. This law was subsequently amended such that the local grid operator took on this obligation instead of the local utility. Table 7.8 shows the tariffs payable to renewable energy generators.

Table 7.8 Tariffs paid to renewable energy generators under the Renewable Energy Sources Act as of 1 January 2002. (Note that the minimum amount paid to wind farms reduces by 1.5% annually for all installations built after 1 January 2002.) (Reproduced from Reference [21])

Technology	Tariff ($€/MWh$)
Small hydro, landfill gas, mines, sewage gas <500 kW	76.7
Biomass <500 kW	101.0
Biomass 500 kW–5 MW	91.0
Biomass >5 MW	86.0
Geothermal <20 MW	89.5
Geothermal >20 MW	71.6
Onshore wind (first five years)	At least 90.0
Onshore wind (after five years)	At least 61.7
Offshore wind (first nine years)	At least 90.0
Offshore wind (after nine years)	At least 61.7
Photovoltaics	481.0

7.6.3 Quota System

Renewables Obligation (RO)

In the UK, the RO places an obligation on all suppliers to source a certain fraction of their demand from renewable energy sources, starting at 3% for 2002/2003 rising to 10% by 2010/2011. If a supplier cannot or will not comply then it can 'buy-out' its obligation by paying (index linked from 2002) £30/MWh (€45/MWh) for the required generation shortfall. This effectively fixes the price for the premium paid to renewable energy generators through a system of certificates known as Renewables Obligation Certificates (ROCs). The proceeds from the buy-out 'pot' are given back to compliant suppliers in proportion to their degree of compliance, i.e. how many ROCs they have bought. This can add extra value to the ROCs. ROCs are a source of revenue to renewable energy generators on top of what they are able to receive through trading their output in the wholesale market.

7.6.4 Carbon Tax

Climate Change Levy

In 2001, the UK government introduced the Climate Change Levy. This is a levy on all energy sold to commercial consumers. Electricity from renewable energy sources is exempt from this levy. The exemption scheme is run using a Levy Exemption Certificate (LEC) scheme. Each certificate has a value of £4.30/MWh or €6.50/MWh (2001 prices), which can be passed on in part or full from the supplier to the generator for the electricity that the supplier supplies to its commercial customers. This levy can be considered as a type of carbon tax, though electricity from nuclear and large scale hydro generation are *not* exempt. The revenue from the tax is recycled in order to reduce employer National Insurance contributions, thus effectively reducing tax on labour to compensate. LECs are also an additional source of revenue over and above the value of the electricity traded by the renewable energy generator competitively.

Eco-Tax Reform

On 1 April 1999, a tax was introduced in Germany on energy use in order to promote energy efficiency and renewable energy sources. The initial value of the tax on electricity was €10.20/MWh, rising by €2.60/MWh annually. As in the UK, this effective carbon tax is used to offset labour tax by reducing statutory pensions insurance paid by employers and employees. Reductions are allowed for certain domestic, commercial and industrial users. Electricity from renewable energy generation sources is exempt from the tax.

Tax Relief

To promote the building of combined heat and power (CHP) plants, wind farms and biogas plants, German investors are able to offset a certain fraction (10%) of the investment costs over 10 years against income tax. This basically gives tax relief on the building of certain renewable energy generation plants.

7.7 Electricity Trading

7.7.1 Introduction

This section will provide an insight into the operation and structure of a modern electricity market, highlighting the differences between a centrally planned and operated system through to a fully deregulated system. This background will stress the challenges faced by renewable energy generation when trying to compete in modern-day markets. In addition, the concept of 'green electricity' trading will be introduced drawing on the twin strands of (a) government legislation to promote the uptake of renewable energy and (b) industry marketing to sell green electricity as a product in demand in the consumer marketplace.

An overview of electricity trading in several countries will be provided but the UK experience will be highlighted as an example of a system that has pioneered electricity deregulation and has seen a market for green electricity develop through this deregulation. The UK example also illustrates the difficulties and evolutionary processes necessary in ensuring that a fully deregulated system is run efficiently and securely.

7.7.2 The UK Electricity Supply Industry (ESI)

The State-Owned Central Electricity Generating Board

Before 1990, the entire UK ESI was in the state sector and was centrally planned and operated by the Central Electricity Generating Board (CEGB). The remit of the CEGB was to provide a secure supply of electricity to its consumers. All generating plant, transmission lines, distribution lines, substations and metering were owned and operated by the CEGB as a monopoly. During the days of the CEGB, electricity was supplied to consumers through the Regional Electricity Companies, which were also responsible for the operation of the distribution network in their area (132 kV and below).

Although this system had worked well for many years, it was a monopoly and severely restricted opportunities for small scale privately-owned generation. During privatization, which was intended to promote competition and drive down costs, the CEGB was broken up into various sectors and sold off to the private sector and new market mechanisms were brought into being.

The Electricity Pool

In 1990, the Pooling and Settlement Agreement (PSA) was introduced and the ESI privatized. The generators were split up between two privatized companies, National Power and Powergen, but nuclear power stations remained under state control. The Regional Electricity Companies (RECs) were individually privatized and given sole licences to supply electricity in their areas, while still retaining responsibility for the distribution network in their areas. The RECs jointly owned the National Grid Company (NGC), which was responsible for the operation of the transmission network (275 and 400 kV) and the pumped-storage facilities in Wales. The NGC also served as the network operator responsible for the scheduling and dispatch of the generation plant.

The Operation of the Pool and Pool Rules

Instead of generation plant being scheduled centrally in an optimum economic dispatch mode reflecting efficiency, operating cost and flexibility, generators now 'bid' into a pool, offering to generate a given output at a given price for the day ahead. The NGC would examine the bids, put them in order of increasing cost and then from this list select the generation plants it would need to meet the demand in a given half-hour period, allowing for transmission constraints and reserve. The most expensive plant required in each half-hour period would set the so-called system marginal price (SMP) for that half-hour. To this price would be added a capacity credit in order to ensure that sufficient capacity over and above maximum demand would always be available for reasons of security of supply. This price was called the pool purchase price (PPP) and was paid to *all* generators selected within the half-hour regardless of whether they had bid less. This pool system is known as a *uniform auction*.

Inflexible plant such as nuclear would commonly bid in at a very low price or even zero to ensure being selected. Note that they would still be paid the PPP even if they had bid zero. Combined cycle gas turbines (CCGTs) with fixed gas contracts would also bid low to ensure being scheduled. The next lowest prices would be operated by the coal fired generators, which were more flexible but with higher operating costs. Finally, open cycle gas turbines and pump storage would bid higher due to higher operating costs and greater flexibility. It can be seen that in general the order of the bids would be in line with the old merit order system but now prices were meant to be driven down due to the competitive bidding nature of the pool.

Figure 7.13 shows a comparison of the average daily half-hourly demand and the PPP for November 1999. It can be seen that the trend in prices throughout the day broadly follows the demand trend during the day. The peaks in prices reflect the more expensive plants that

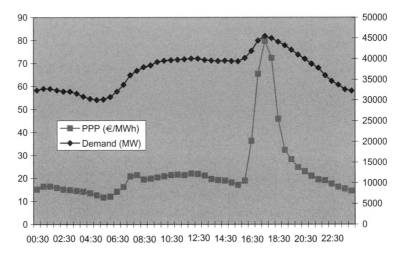

Figure 7.13   NGC average daily half-hourly demand and pool purchase price for November 1999

need to be chosen by the NGC to meet the peaks in demand. The PPP was then 'uplifted' slightly to cover the difference in the total cost incurred by the NGC in balancing generation and demand and what was paid out to generators through the PPP. This uplifted price was known as the pool selling price (PSP). The PSP is then charged through to suppliers for the amount of electricity supplied to their customers.

All generation and demand is metered. Large generators and consumers are metered every half-hour. Small consumers are metered monthly, quarterly or six-monthly. All meter readings are processed and made available to the system operator (SO), who works out who should be paid, what amount and who owes what amount. Additionally, charges are made for the use of the transmission and distribution networks. The process of allocating the costs and payments is known as *settlement*.

Hedging

It can be seen from Figure 7.13 that there is a wide variation in prices during the day. The prices shown are monthly averages. It was quite possible to get price fluctuations at times of peak demand (around 5:30 pm), well over £100/MWh (€150/MWh). In certain severe weather conditions this could go over £250/MWh (€375/MWh). For a supplier with a peak 'tea-time' demand of 2 GW (2000 MW) this could mean half-hourly cost fluctuations approaching £0.25 million (€0.38 million)! Most suppliers operate to quite tight margins and charge a fixed price to their customers, so a sudden increase in pool prices over a cold period could potentially bankrupt a supplier. On the other hand, generators are paying a fixed price for their fuel and similarly a dip in prices during a prolonged warm period could also spell disaster for their cash flow. Soon after the introduction of the PSA, so-called 'Contracts for Differences' came into existence. These 'financial instruments' gave generators and suppliers the ability to fix the price of their generation or demand for a given period of time, typically a month or more, but also up to several years ahead. The contracts did this by *hedging* against changes in the pool prices.

Typically a supplier would forecast its demand profile by the half-hour throughout a month, taking into account changes due the seasons, changes due to expected customer losses/gains, etc. This would be the *hedged volume*. The supplier would present this profile to a third party broker, which could be a financial institution. This broker would then offer a fixed price for the contract, reflecting how volatile the broker expected the prices to be in the next month. The supplier would still have to pay the half-hourly PSP for demand but at the same time if the PSP went above the fixed price, the broker would pay the supplier for the difference between the PSP and fixed price multiplied by the hedged volume. On the other hand, if the PSP went below the fixed price for any half-hour, the supplier would pay the broker for the difference between the fixed price and the PSP multiplied by the hedged volume. In this way the supplier was effectively paying a fixed price for its hedged volume. Obviously, if the hedged volume were an underforecast, the supplier would be *exposed* for the difference between the forecast and actual demand. If the forecast were too high, then the supplier might be paying over the odds for the contract, though it could be possible to make a profit with some speculation. Generators could fix the price they were paid for a hedged volume of generation in exactly the same way.

Eventually, bespoke contracts became 'off-the-shelf' contracts with standard terms and conditions and standard contract types, for example for base load or peaking. These were known as Electricity Forward Agreements (EFAs). These could be bought and sold on power exchanges. In this way, suppliers and generators could buy and sell in a *forward trading* market. The importance of such contracts became apparent when the independent supplier Independent Energy lost its credit rating and could not obtain an EFA, finally becoming bankrupt due to excessively high generation prices in September 2000.

Deregulation

Immediately following privatization, consumers could still only buy their electricity from their local REC. However, in 1992, consumers with an average demand of greater than 1 MW could choose a supplier outside their local area. In 1994, the demand threshold was reduced to 100 kW. Eventually, in 1999, all consumers were able to choose their supplier.

At the same time, small independent generators and *second tier* suppliers entered the market, breaking the monopoly of National Power, Powergen, Nuclear Electric and the RECs (which latterly became known as Public Electricity Suppliers or PESs). This gave an opportunity for renewable energy generators and specialist renewable energy suppliers (such as unit[e] and Ecotricity) to enter the market. In addition, the PESs have been unbundled so that supply, distribution, metering and meter reading have all become separate businesses. Over time the larger suppliers like National Power and Powergen have become fragmented and taken over so that they bear little resemblance to the companies they were at privatization. In addition, the National Grid Company, still the system (and transmission line) operator, became an independent company.

The New Electricity Trading Arrangements (NETA)

The electricity pool system was felt to have played its part in the successful deregulation and opening up of the ESI. However, criticisms were made of the way in which pool prices could be manipulated by a few key generators bidding into the pool. In fact, it was considered that this 'rigging' of pool prices was keeping wholesale electricity prices artificially high. In addition, it was felt that generators and suppliers should be free to enter into *bilateral* agreements instead of also buying from the pool, albeit with a bilateral hedging contract sitting on top. Furthermore, it was seen as beneficial that third parties could enter the market to trade in physical electricity much as they had been doing in CFDs and EFAs. For this reason, the New Electricity Trading Arrangements (NETA) were introduced in March 2001.

Buying and Selling Electricity Under NETA

Generators and suppliers (above a certain size) are obliged to notify their position for each half-hour in terms of generation and demand for the day ahead to the National Grid as system operator. This allows the NGC to carry out its balancing of supply and demand much as before. Generators now *self-dispatch* rather than wait to be dispatched by the NGC, although they can alter their output when requested to do so or if they participate in the balancing market (see below). Generators still self-dispatch in a very similar order to the merit order under the CEGB regime, but this time the order of dispatch is dictated solely by price. In addition, suppliers and generators notify what contracts they have struck to the NETA central

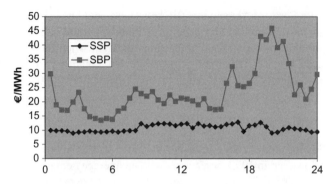

Figure 7.14   Average daily half-hour system sell and system buy prices for March 2002

systems. These contracts must be notified at least 1 hour[1] ahead of actual time, known as *gate closure*. Instead of suppliers dealing in CFDs and EFAs to hedge an agreed volume against pool prices, they now trade in *physical* volumes at a fixed price. Any difference between the physical volume contracted and actual volume generated or supplied is *cashed-out* at prices set within the balancing market.

Suppliers and generators typically strike contracts on different timescales. A month or more ahead of actual delivery they will strike *bilateral contracts*, which are straight contracts between two parties for a given volume of electricity at a given price. This accounts for 90% of the volume of wholesale electricity bought and sold. Suppliers and generators may not be able to predict their demand or output exactly months ahead, so around a day ahead they buy and sell chunks of power on a *power exchange*. This is much like a stock exchange for power, where 'chunks' of power are bought and sold anonymously. This can be done up to gate closure, as mentioned above. Around 5% of electricity is traded in this way. The remaining 5% is traded through the cash-out or imbalance market.

The Imbalance Market

If a supplier's actual demand in a half-hour is higher than it has contracted for ahead of gate closure, then the supplier must pay the system operator to top-up this deficit at the system buy price (SBP). If the supplier's actual demand is lower than it has contracted for it is paid the system sell price (SSP) for the spill excess. This *imbalance market* is therefore known generally as a *top-up and spill market*. The principle is identical for a generator, whereby it will need to pay the system buy price for any deficit in the contracted generation output and will be paid the system sell price for any generation excess above the contracted output. The system sell price is in general lower than the average bilateral contract price between the supplier and generator. The system buy price is in general higher than the average bilateral contract price between the supplier and generator.

Figure 7.14 shows the SSP and SBP for each half-hour average over the month March 2002. The SSP is around £10.70/MWh (€16.10/MWh) for this month, and the SBP is £23.70

---

[1] This was initially 3.5 hours ahead at NETA 'Go-Live' but was reduced to 1 hour ahead in July 2002.

(€35.60/MWh), though the SBP shows rather more variation during the day, crudely reflecting the shape of the national demand curve. The typical contract price for this month would be around £15/MWh (€23/MWh). It can be seen that there is an incentive for suppliers and generators to get their forecast as close as possible to their actual position due to these non-symmetrical prices.

The Balancing Market

The system operator, National Grid, needs to give orders to generation plant to change their output in order to match supply and demand, just as in the days of the CEGB and the pool. This is done using another market known as the *balancing market*. Ahead of delivery, flexible generators (and to a lesser extent flexible suppliers) can place bids to reduce power output or offers to generate extra output. In the case of a supplier it would make a bid to increase demand or an offer to reduce demand. After gate closure, the NGC examines the bids and offers and accepts those it needs to balance the system. Suppliers and generators pay their accepted bids and are paid their accepted offers. This type of market is therefore known as a *bid-price auction*. Simplistically, the value of the bids sets the system sell price and the value of the offers sets the system buy price, though there are other costs and adjustments that feed into these prices. The total cost of balancing the system in general is slightly less than the balance of the proceeds for the SSP and SBP. This is due to the fact that two generators may be out of balance separately and pay for this, but together their imbalances cancel out. The excess is reallocated to suppliers and generators in proportion to their total generation output or total demand as appropriate.

The British Electricity Transmission and Trading Arrangements (BETTA)

Until April 2005, the electricity wholesale market in Scotland operated differently to that in England and Wales, though there were similarities in terms of 'top-up' and 'spill' prices. After this date, the two markets were merged to give a version of NETA operating across Great Britain. At the same time, the transmission network in Scotland, which had been operated by the two Scottish generation and supply companies, became independently operated by National Grid, who consequently became responsible for the entire Great Britain transmission network. These new arrangements became known as the British Electricity Transmission and Trading Arrangements (BETTA).

7.7.3 Competitive Wholesale Markets in Other Countries

Introduction

The England and Wales pool-based wholesale electricity market was ground-breaking when it was introduced. Since its introduction, other countries have introduced wholesale electricity markets based on power pools, bilateral forwards and futures markets and balancing markets.

Broadly speaking, the wholesale price of electricity can be set in advance (*ex ante*), in real time or after the event (*ex post*). Ex ante prices can typically be set at the *day-ahead* stage.

Clearly, it is not possible to know exactly what a generator will produce or a consumer demand ahead of time and such markets with ex ante prices normally require some kind of balancing market to settle the differences between actual and projected positions. The system operator calls upon bids and offers within the balancing market to balance the system and prices in such markets are set ex post as they reflect the actual cost of balancing the system. Ex post markets reflect the actual cost of operating a system and thus require no secondary balancing market. Normally, a system operator will publish indicative day-ahead prices so that market players can adjust their position to maximize revenue/minimize cost.

The Scandinavian Nord Pool

Sweden, Norway, Finland and parts of Denmark trade wholesale electricity through the Nord Pool. The Nord Pool operates a day-ahead *spot market*, a forwards market and a futures market. Most electricity is traded through bilateral contracts and trading in the Nord Pool is voluntary. Market participants submit bids and offers to the spot market (taking into account bilateral contract commitments) for each hour of the day ahead. The Nord Pool set ex ante prices from the intersection of the aggregate supply and demand curves. An example of this is shown in Figure 7.15.

This figure shows a typical set of supply and demand curves. As prices increase demand will tend to go down as consumers cannot or will not pay the higher prices. On the other hand, an ever-larger number of generators are able to provide power as the price they receive increases, thus making it attractive for more expensive generation to run. In this example, the equilibrium point is reached at around 310 units of power, setting a spot market price of around 40 units.

By 13:30 on the day ahead, market participants have to confirm their bids and offers, which then become firm. The system operators in the participating countries operate balancing markets where generators and suppliers can submit bids and offers to adjust their output. An ex post balancing market price is then set for each hour by the marginal bid or offer used. Any imbalances are then settled for each participant at this price.

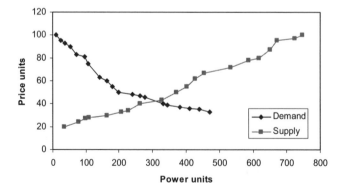

Figure 7.15   Representative supply and demand curves for the Nord Pool

Australia

Between 1994 and 1997, in the state of Victoria the VicPool market used to operate. This was very similar to the England and Wales pool. The system operator set ex post system marginal prices for each half-hour respectively. Generation and demand side bids and offers on a rolling seven-day cycle could be submitted. Participants could alter their bids and offers up to a day ahead based on indicative prices released by the system operator. The VicPool, like the England and Wales pool, was a mandatory market. Australia now has a mandatory National Electricity Market (NEM). Under this market generators and suppliers submit bids and offers a day ahead. The market operator NEMMCO then runs a scheduling program to calculate the ex ante system marginal price five minutes ahead of time for the next five minutes. Deviations from the projected schedule are met by ancillary services. The effective trading period is a half-hour and settlement prices are calculated as the time-weighted average of the five minutes over the half-hour.

New Zealand

An optional spot market (NZEM) with ex post pricing is operated in New Zealand. Participants submit bids and offers at the day-ahead stage. Forecast prices are issued at 15:00 on the day ahead. Participants can re-bid prices and volumes up to two hours ahead of dispatch. Bilateral contracts must be notified to the system operator and any deviations are settled at prices emerging from the NZEM. The majority of participants trade through the NZEM.

Argentina

Generators and large consumers have the option of participating in a spot market. In addition, they can enter into bilateral contracts. Suppliers to smaller consumers buy electricity at regulated seasonal prices or directly from generators under bilateral contracts. Seasonal prices are set by the government every six months based on spot prices. The spot market determines hourly ex ante prices for the day ahead. The prices can be modified up to an hour ahead if there are significant changes to the supply/demand balance. Generators bid in their fuel costs every six months subject to a price cap. Hydro generators determine the value of water every six months. The market operator CAMMESA calculates marginal generation costs using predefined algorithms and the bid fuel/water prices.

California

California operates an optional power exchange (PX) to set ex ante prices for the day ahead. Generators and suppliers submit day-ahead bids and offers based on an iterative auction where the participants can revise their bids and offers up to five times. After each iteration the PX publishes indicative prices and whether or not participants' bids have been accepted. The PX submits the final day-ahead schedule to the independent system operator (ISO). At the same time participants who have not participated in the PX, either because they have entered into bilateral contracts or have traded on different power exchanges, submit their positions to the ISO through scheduling coordinators (SCs). Revised schedules can be submitted up to an hour ahead of real time. Balancing is carried out by the ISO via a balancing market based on hour-ahead adjustment bids. An ex post balancing price is then set to settle participants' imbalances.

Deregulating Electricity Markets in Europe

Within the EU, initially Norway, Sweden, Finland and the UK liberalized their electricity markets to promote competition and lower prices. In 1996, the European Commission issued a directive for other EU countries gradually to liberalize their electricity markets. As a result, other EU countries have started a programme of liberalization, though some are more advanced than others.

A couple of European Union countries are adopting the so-called *single-buyer model* (SBM). Under this market mechanism, all power producers sell their output to a single buyer (SB). The SB is also the sole authority allowed to sell electricity to consumers. Producers tender for long term Power Purchase Agreements (PPAs) with the SB. This mechanism is favoured by France and Greece. Other EU countries are introducing market systems with a combination of power pools, power exchanges, bilateral trading and balancing markets.

7.7.4 The Value of Renewable Energy in a Competitive Wholesale Market

Introduction

This section deals with the valuing of renewable energy generation when competing against other forms of generation in an electricity market. The following section deals with the issue of the 'green value' of the electricity. There are two basic differences between renewable energy generation and 'conventional' generation with regard to trading in an electricity market. These are:

- the relatively small size of renewable energy generation units;
- the variability or lack of controllability for most renewable energy generation units.

The first issue indicates that such generators tend to have less *leverage* in a competitive market. In order to make a reasonable profit, electricity suppliers have relatively large customer bases and may demand thousands of MW. Renewable energy generators may only produce tens of MW (though some of the proposed offshore wind farms may produce significantly more than this). Many of the power exchanges, for example, do not accept bids or offers of less than 1 MW and even if a renewable energy generator is able to trade in 1 MW blocks, this market is likely to be relatively *illiquid*, that is to say, trading blocks of this size are not easily sold in the market. These drawbacks will tend to restrict the value of the renewably generated electricity.

The second issue is particularly important in a market that sets ex ante prices and has a balancing market for top-up and spill. Wind power generation is variable, relying on changes in the wind, which can be forecast to a degree but with diminishing accuracy the further ahead one looks. Hydro generation, without storage, is dependent on river flow rate, which in turn relies on rainfall. Reservoir storage will help to mitigate fluctuations, but a prolonged drought will cause generation output to cease. Biomass generation, though controllable, is dependent on the biomass feedstock, which may be seasonally dependent. Photovoltaic power can be reasonably predictable in a warm climate with little cloud, but the movement of clouds can cause significant fluctuations in output. Wave power generation is reliant on wind to create waves. Changes in the energy in waves tend to be smoother than changes in wind energy as the waves tend to 'integrate' the energy that the wind imparts. Tidal power relies on the relative phases of the moon and sun and as such is very changeable on a daily basis but is reasonably predictable. Changes in wind speed/direction and pressure can modify the expected tidal range.

The topic of variability was discussed at length in Chapters 2 and 3 and the summary above highlights that all renewable energy sources have issues relating to variability to varying degrees and on different timescales. The ability to forecast changes also varies from one renewable energy generation source to another. If an intermittent renewable energy generator contracts bilaterally for a given amount of energy with another party, e.g. a supplier, any difference between what the generator is contracted to supply and what the generator actually supplies will be 'cashed-out' at top-up and spill prices emerging from a balancing market. These prices are generally unfavourable compared with average bilateral prices. In this type of market, renewable generators that cannot accurately predict their output are disadvantaged.

Case Study: The Impact on a Small Hydro Generator of an Electricity Market with Bilateral Trading and a Balancing Market

A small hydro generator wishes to trade its power output into an hourly electricity market where participants are able to trade bilaterally and cash-out imbalances in a balancing market at asymmetric top-up and spill prices, i.e. top-up and spill are charged/paid at different rates reflecting bids and offers from flexible generators/consumers who are able to adjust their output at short notice if called upon by the system operator. The hydro generator has entered into a day-ahead contract with a supplier to provide a fixed output of 1.9 MW every hour over a 24 hour period at a price of €15/MWh. What will be the generator's average revenue in €/MWh over this 24 hour period?

To answer this question, the actual hourly output of the hydro generator and the hourly top-up and spill prices over the 24 hour period need to be known. Table 7.9 illustrates the required calculation. It can be seen that the hydro generator's output varies throughout the day above and below the 1.9 MW value for which it was contracted. As far as the supplier is concerned, the 1.9 MW is supplied as contracted and the supplier will pay the generator a total of €684.00 for the energy supplied during the day. This contract is notified in advance to the market operator. However, the hydro generator's output is metered hourly and the market operator can see that output does not match the contract at each hour. The generator is thus obliged to make up shortfalls at the top-up unit price (TUP) and is paid the spill unit price (SUP) for any excess. In this market, as with most markets having asymmetric balancing prices, the SUP is rather less than the prevailing market price and the TUP is rather more than the prevailing market price. The net value of energy deficits and surpluses then has a value of cost to the hydro generator of €53.40. If this is added to the

Table 7.9 Illustration of the value of a hydro generation contract at €15/MWh with imbalance cashed-out at spill unit prices (SUPs) and top-up unit prices (TUPs)

Hour	Actual (MWh)	Contracted (MWh)	Contract value (€)	SUP (€/MWh)	TUP (€/MWh)	Imbalance cost (€)
1	1.0	1.9	28.50	11.00	23.00	−20.70
2	1.3	1.9	28.50	11.00	21.00	−12.60
3	1.5	1.9	28.50	9.00	20.00	−8.00
4	2.0	1.9	28.50	8.00	23.00	0.80
5	2.0	1.9	28.50	10.00	28.00	1.00
6	2.0	1.9	28.50	12.00	30.00	1.20
7	2.5	1.9	28.50	12.00	35.00	7.20
8	2.5	1.9	28.50	13.00	33.00	7.80
9	2.5	1.9	28.50	13.00	35.00	7.80
10	2.6	1.9	28.50	13.00	37.00	9.10
11	2.8	1.9	28.50	12.00	33.00	10.80
12	2.7	1.9	28.50	11.00	30.00	8.80
13	2.6	1.9	28.50	10.00	28.00	7.00
14	2.5	1.9	28.50	9.00	27.00	5.40
15	2.3	1.9	28.50	9.00	25.00	3.60
16	2.0	1.9	28.50	8.00	24.00	0.80
17	1.5	1.9	28.50	8.00	23.00	−9.20
18	1.4	1.9	28.50	7.00	24.00	−12.00
19	1.5	1.9	28.50	7.00	23.00	−9.20
20	1.6	1.9	28.50	8.00	24.00	−7.20
21	1.5	1.9	28.50	9.00	23.00	−9.20
22	1.3	1.9	28.50	9.00	21.00	−12.60
23	1.2	1.9	28.50	8.00	20.00	−14.00
24	1.4	1.9	28.50	9.00	20.00	−10.00
	46.2	45.6	684.00			−53.40

revenue from the generator's contract, the generator makes a total of €630.60 for the day. The total amount of energy for which the generator was contracted was 46.2 MWh. Thus the net value of the generator's energy works out to be €13.65. Therefore, the penalty for the generator's imprecise forecast of power output a day ahead is around 9% of the bilateral contract price.

Trading Renewable Energy in an Electricity Market with Ex Post Pricing

If the same hydro generator were trading in a market with ex post pricing, for example a pool system like the former England and Wales pool, it would not be exposed to unfavourable imbalance prices. The major difference between such a mandatory pool system and a system with bilateral trading and a balancing market is that the former spreads the cost of balancing the entire system on all participants, whereas the latter targets the cost of balancing on those participants who are out of balance compared with their contracted position. In general, mandatory pool prices are less volatile than balancing market prices, as all power must be traded through it, so the price penalty to intermittent renewable generators is less. However, if a renewable generator wished to hedge its position in order to receive a fixed price for output, it would still be exposed to pool prices for any generation not covered by the hedging contract. If pool prices suddenly decreased during a period due to a surplus of generating capacity, the renewable generator would then lose out in this case, particularly if it had underforecast significantly and thus underhedged.

Marketing Green Electricity

Though government legislation is one way to encourage renewable energy generation, it is not the only way. The opening up of electricity markets has given the opportunity for new and existing suppliers to offer green tariffs to consumers. Green electricity is marketed as a 'premium product' for consumers with an environmental conscience. Generally, a premium is charged for the green tariff. This premium is either paid directly to the renewable energy generator or used to support new renewable energy generation. The take-up of these tariffs has varied from country to country. In the UK, for instance, market research suggested 25% of domestic electricity customers, representing 5.7 million households, would be interested in a green electricity tariff, even if this means paying a little more than the lowest prices to ensure their electricity comes from renewable sources. In reality, however, less than 1% of households in the UK has switched to a green tariff, while the Dutch and American markets have achieved a figure of 2%. It seems that the premium that consumers are willing to pay may be quite small. In addition, the freedom to change electricity suppliers is a relatively new concept, as is the idea of buying green electricity, so it may be too early to draw firm conclusions about the success or otherwise of marketing green electricity.

References

[1] Danish Energy Agency, 'Technology data for electricity and heat generating plants', 2004.
[2] International Energy Agency, *Renewables for Power Generation, Status and Prospects*, OECD, Paris, 2003.
[3] IEE Power Division, *Economics of Embedded Generation*, Institution of Electrical Engineers, London, 1998.
[4] Hohmeyer, O. *Social Costs of Energy Consumption*, Springer-Verlag, Berlin, 1998.
[5] 'Externalities of energy (ExternE), European Commission, 1995–99', EUR 16520-25 EN; also 'External costs. EC summary document on the above', published 2003, EUR 20198. The EU page http://euaropa.eu.int/comm/research/energy/gp/gp_pubs_en.html has a link to these and other relevant material.
[6] *Power generation and the environment*. Roberts, Liss and Saunders, Oxford University Press, 1990.
[7] Lockwood, B. 'The social costs of electricity generation', CSERGE, University of East Anglia, 1992.
[8] A full description may be found at http://europa.eu.int/comm/environment/climat/emission.htm and on the DEFRA site at http://www.defra.gov.uk/environment/climatechange/trading/eu/.
[9] Wenger, H. and Hoff, T. 'The value of PV in the distribution system – the German grid-support project', PG&E R&D Report 007.5-94.15.
[10] 'REVALUE – the value of renewable electricity', European Commission Project JOR3-CT96-0093.
[11]Laughton, M. 'Implications of renewable energy in electricity supply', IMechE Seminar on *Power Generation by Renewables*, Professional Engineering Publishing, Bury St Edmunds, 2000.
[12] National Grid, 'National grid and distributed generation – facilitating the future', 2001; www.nationalgrid.com/uk/social&environment.
[13] The Transmission Issues Working Group Final Report, DTI, 2003.
[14] 'Charge change prompts green fears', *Power UK*, Issue 119, January 2004.

[15] South Western Electricity plc., 'Interaction of Delabole wind farm and South Western Electricity's distribution system', ETSU Report W/33/00266/REP, 1994.

[16] www.dti.gov.uk/energy/renewables/tiwgreport.pdf.

[17] Milborrow, D.J. 'Capacity credits – clarifying the issues', in British Wind Energy Association 18th Annual Conference, Exeter, 25–27 September 1996, MEP Ltd, London, 1996.

[18] Ilex Energy Consulting Ltd and UMIST, 'Quantifying the system costs of additional renewables in 2020' (The 'SCAR' report), Report commissioned by DTI.

[19] Dale, L., Milborrow, D., Slark and Strbac, G. 'A shift to wind is not unfeasible', *Power UK*, 2003, 109, 17–25; on BWEA website www.bwea.com.

[20] Harrison, L. 'Wind economics set to beat gas in Ireland', *Windpower Monthly*, December 2004.

[21] Ruchser, M. 'ENER-IURE Project Phase III – analysis of the legislation regarding renewable energy sources in the EU member states', Report concerning electricity in Germany, 2001.

8

The Future – Towards a Sustainable Electricity Supply System

8.1 Introduction

We live in a world of enormous divides in terms of wealth, and these are reflected in major disparities of energy consumption per capita. An average American consumes a massive 13 000 kWh per year of electricity while in the underdeveloped regions of Africa the average lies between 50 and 100 kWh. In fact, nearly two billion people have no access to electricity at all. In terms of social justice and for more pragmatic considerations such as reduction in tension between the haves and the have-nots, it is essential that this imbalance is addressed.

A universal global electrification scenario that allows all of the world's population to attain the current average per capita electricity consumption of just over 3,000 kWhrs per annum, would require a massive increase in electricity generation capacity. Is it feasible to achieve this using traditional fuels or nuclear power? Assuming that climate change is indeed happening, the first option would be an environmental disaster and it would substantially shorten the period over which the limited oil and gas reserves would be available. The second option might appear to be environmentally benign but would place unsustainable demand on uranium supplies and give rise to additional quantities of nuclear waste that would exacerbate present concerns regarding safe disposal.

The only sane alternative is a major increase in the exploitation of renewable energy sources, leading eventually to a completely sustainable electricity supply system. Effecting a transition from our present rapacious consumption of finite resources is of course a massive challenge, but one that has to be met face on.

Renewable resources have a range of advantages. They are:

Renewable Energy in Power Systems   Leon Freris and David Infield
© 2008 John Wiley & Sons, Ltd

- nondepletable and will be there as along as the sun is shining;
- indigenous, hence reducing dependency on imported fuel from other countries with unstable regimes;
- virtually nonpolluting, with some small emissions produced during manufacture and end-of-life disposal;
- diverse and complementary in their time dependence;
- generally small and geographically distributed and can often be located near to the demand, reducing transmission and distribution losses;
- robust in system terms because of the very large numbers of individual generators and the statistical robustness of such a collection compared to centralized generation;
- particularly suited to the needs of developing countries, where systems based on renewable energy technology can be simpler to manufacture than traditional fossil fuelled or nuclear power stations.

However, renewable energy sources involve relatively new and underdeveloped technologies that have the following disadvantages:

- In general, the cost of electricity from some, but not all, such sources is at present higher than that from conventional energy sources. This comparison, however depends importantly on the inclusion or exclusion of external costs;
- Except for biomass based generation, they are nondispatchable;
- Their distributed nature may require restructuring of the electricity supply infrastructure.

This last area may be of particular interest to readers of this book who are now in a position to understand the key technical features required of future electricity supply systems. These will be explored in more detail in the remainder of this chapter alongside projections for the expected evolution of the individual renewable energy technologies.

8.2 The Future of Wind Power

8.2.1 Large Wind Turbines

Recent progress in wind turbine technology has been breathtaking. At the time of writing 5 MW wind turbines are commercially available. Ten years ago, the largest wind turbine on the market was typically an 800 kW machine with most manufacturers offering 600 kW units. At present, most wind turbines are fitted with rotors of 60–90 m compared to the 35–45 m of the mid-1990s. Due to economies of scale, larger machines generate electricity at a lower cost per kW h , particularly in offshore applications.

This increase in size has been evolutionary in nature with successive but incremental scaling up of proven designs, allowing manufacturers to progressively improve their understanding of and confidence in the design process and key issues such as fatigue. However, is this the best recipe for future expansion? Might not step change innovations provide a faster path to lower cost larger machines? And is there a limit to the size of wind turbines imposed by material strength limitations? Since the emergence of the first commercial wind turbine in the mid-seventies experts have been predicting definite limits to turbine size. All these limits have now been exceeded, pointing up the hazardous nature of such predictions.

Figure 8.1 Artist's impression of the ICORASS 10 MW wind turbine. (Reproduced from Polymarin)

Figure 8.1 shows an artist's impression of a radical design for a 10 MW wind turbine proposed by a consortium led by Dutch rotor blade maker Polymarin Composites. This turbine known as ICORASS, which stands for integral composite offshore rotor active speed stall-control has a 160 m two-blade rotor and is intended for offshore applications where foundation and installation costs do not scale linearly with turbine power, providing cost savings for larger unit sizes. The design is highly robust to provide exceptional reliability and availability as required for offshore application where access and maintenance is difficult. Key features of this particular design are the integration of a direct drive alternator into the hub, a fixed blade angle with stall power control and the use of a permanent magnet alternator for speed reduction (braking) above rated wind speed to ensure fixed power delivery, and lastly a downwind free yawing rotor to remove the need for an active yaw drive system.

It is too early to say whether such advanced design concepts are likely to be commercially successful, but it is clear that future turbines will be significantly larger than today's, especially where offshore application is concerned, and that wider use will be made of advanced composites. Alongside innovations in construction, increased use will be made of sophisticated condition monitoring and preventative maintenance systems.

8.2.2 Offshore Wind Farm Development

An offshore installation has advantages over and above the reduction in turbine costs from the resulting economies of scale. The marine environment is windier resulting in an increased energy yield, and the winds are more persistent and less turbulent making system-wide integration easier. Very large wind turbines that might cause an unacceptable visual intrusion onshore may well be acceptable if sited away from the shore. Additionally, the transportation of exceptionally large or heavy items is easier over water than over land where roads may need widening of reinforcement. Importantly, planning approval may be easier to obtain for offshore schemes due to their reduced visual and noise impact. Such advantages will become increasingly important as the acceptable onshore sites are progressively used up. At present, offshore wind energy is more expensive than on shore, primarily due to significantly higher foundation, installation and electrical connection costs. Complex and costly underwater cable systems are required to link the turbines together and to make the connection to shore. In some cases an offshore substation may be required. However, as more off-shore plant is built, the technology will advance to deal with unanticipated problems and to reduce transportation and installation costs. For example, while lightning tends to strike onshore wind turbines at the blade end, in offshore applications it was observed that strikes are more frequent at the centre of the blade. Blades are now redesigned to cope with such occurrences. To speed up installation dedicated vessels are being developed and built to deliver and install the towers nacelles and blades.

One of the major problems is the accessibility of turbines for maintenance. This is only feasible by sea during relatively calm conditions, so the larger off-shore wind turbines are sometimes fitted with platforms on which personnel can be lowered from a helicopter. These procedures are costly so designers are under pressure to develop turbines with exceptional reliability and long maintenance intervals. As already mentioned, sophisticated condition monitoring systems continuously assessing the turbine performance and fatigue state will become increasingly common. Such systems were too expensive for onshore turbines of 1 to 2 MW but for multimegawatt units the systems represent a much smaller fraction of the turbine capital cost; moreover the costs of instrumentation are falling making such approaches more cost effective.

Electrical Integration

The growth of significant offshore wind capacity has raised grid integration issues to a new prominence. Increasingly large amounts of electricity will be feeding into national networks at points not specifically designed for such infeeds. Consequently, significant network rein-forcement will often be required. In addition these time varying infeeds must be integrated into grid management systems not previously required to cope with such non-dispatchable sources. The costs of expected system adaptation will be significant and new market mecha-nisms may well be needed to ensure that the required infrastructural investments are made in a timely manner.

Today's onshore wind turbines typically generate at around 700 V, which is commonly stepped up to ~35 kV by a transformer in the nacelle or the tower base. The first offshore schemes have tended to use a similar arrangement, although it is anticipated that as wind farms become larger with longer distances between turbines and increased distance to shore, the voltage will have to increase.

In the Danish 40 MW Middelgrunden wind farm a voltage level of 30 kV was chosen for the collection of power between the 20 turbines and the short 3 km connection to shore. In contrast for the 160 MW Horns Rev wind farm, the collection voltage is 36 kV but the voltage is then raised to 150 kV at an offshore transformer substation before transmission over 15 km to shore. In future large offshore schemes AC transmission may severely limit the distance over which electrical power may be transmitted by undersea cable.

DC Transmission [1], [2]

AC cables possess large capacitance resulting in large reactive power currents which are proportional to cable length. Therefore the useful power that can be transmitted reduces as length and voltage increase, with viable AC cable lengths limited to only a few tens of kilometres. Cables when run with DC have no length limit. Additionally, they are cheaper and less lossy. DC schemes also provide a better level of reliability if a single cable is lost.

In the longer term, as wind farm sizes grow, perhaps to several GW, and the distance to shore increases, it is likely that DC transmission may be used to link the wind farms to the grid network. The cost of a DC line to transmit a certain power is considerably lower than that of the AC equivalent. However, generators are most efficient when designed to provide AC, and AC is unbeatable when used to distribute energy to consumers at different voltage levels. Hence DC transmission requires conversion from AC to DC and from DC to AC at the extremities of the DC line. This is achieved through power electronic converters.

DC transmission is used extensively at present in power networks when bulk power has to be transmitted over long distance overhead lines or medium distance underground or undersea cables. It can be shown that if the power to be transmitted is large enough and the overhead or cable line is above a critical length, the savings in transmission line costs more than compensates for the extra expense of the converters. Because of the DC based coupling of the AC networks, a DC link is asynchronous i.e. the two linked AC systems do not have to operate at the same frequency. This was illustrated in Figure 4.49 where a gearless wind generator is connected to the network through a DC link. In conventional HVDC (high voltage DC) links the converters used are the line-commutated converter of Figure 4.26. This technology has a proven track record but also suffers from the disadvantages listed in Section 4.5.5.

An HVDC link can be realized using the voltage source converters (VSC) described in Section 4.5.6 rather than line-commutated converters. This technology has only become possible through recent advances in high power IGBTs. Unlike line-commutated converters, VSCs have no need for an AC source commutation voltage, are capable of independently controlling the active and reactive power and inject negligible amounts of harmonics into the AC network. Another advantage of the VSC based HVDC is that it can be built on a modular basis using standardized units to build up a converter station. It is because of these characteristics that utilities are usually in favour of VSC based HVDC technology. Compared to the AC option the VSC-HVDC would be substantially more expensive in near to shore wind farm schemes but would become more favourable with increasing scheme sizes and distances.

Figure 8.2 shows the key parts of a VSC based HVDC terminal. A mirror image of this terminal is connected at the other end of the transmission cable to complete the system. The heart of a terminal consists of a three phase IGBT based bridge, as shown in Figure 4.35. Because of the limited voltage and current rating of the largest available IGBTs each of the six 'valves' consists of a number of parallel–series-connected IGBTs to increase the power handling capability of the terminal. IGBT chips and diode chips are connected in parallel in a submodule. A StakPak IGBT has two, three, four or six submodules. The number of submodules is based on the required current rating of the application. For high

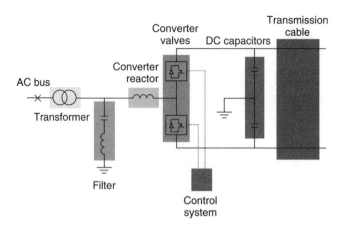

Figure 8.2 Schematic of a VSC based HVDC system. (Reproduced with permission of Asea Brown Boveri Ltd)

power ratings requiring a high DC voltage, more than one hundred StsckPacks are connected in series.

If the desired DC voltage does not match the AC system voltage, a normal AC transformer may be used in addition to the converter series reactor. The converter has a PWM switching frequency of 2 kHz. Since there are only high frequency harmonics generated on the AC side, filtering can be done using a simple high pass filter. On the DC side there is a capacitor that also serves as a DC filter. The two pole configuration shown in Figure 8.2 where the middle of the two pole DC system is earthed, enhances transmission reliability. In the case where a fault occurs in one pole, the other can provisionally continue transmission with an earth return.

This type of reliability is not possible in a three-phase AC system. After a transient loss of the AC supply, the asynchronous nature of the HVDC link allows 'black-starting' by controlling the frequency from zero to the required wind turbine frequency, while ramping up the wind farm voltage.

The VSC based HVDC technology is advancing fast. Manufacturers will soon be able to offer such links rated at 1 GW and operating at ±300 kV on the DC side. This will substantially extend the off-shore range and size of wind farms that can be potentially exploited. Whether such schemes will be realized in the future depends on the shortage of near-shore sites and on the technical success and reliability of the present offshore schemes, which are operating or are under construction.

Other Demands on Large Offshore Wind Farms

Due to the economies of scale, future offshore wind farms are likely to be an order of magnitude larger than the present onshore schemes. The rating of such wind farms will be comparable to that of other traditional generating plant on the grid. As a consequence, they will be required to perform functions not demanded from present onshore wind farms.[1] Such functions will include:

- control of active and reactive power;
- system frequency control;
- local grid voltage control;
- operation under abnormal conditions such as grid faults;
- contribution to network stability.

These requirements will demand significantly improved sophistication of wind turbine and wind farm control functions. Although the majority of the current multimegawatt wind turbines operate at variable speed using a doubly fed asynchronous generator some experts believe that this arrangement is not optimum if stricter utility requirements have to be satisfied. Compared with full power converters such as those in gearless wind turbines, doubly fed arrangements are more problematic because during fault situations they can inject large peak currents into the system.

If the ICORASS design is a true representation of future wind turbines, its permanent magnet synchronous machine can only be interfaced to the grid through a full power converter which, in addition to the utility requirements listed above, will have the added task of braking the wind turbine above rated wind speeds to enforce stall regulation power control. Future wind turbine designers will not be short of challenges!

A Proposed European DC Supergrid

The VSC based HVDC technology can be adapted to operate with several terminals on the DC side so that power can be injected or extracted at will at each terminal. This proposed mode of operation has encouraged feasibility studies for a European DC supergrid. Because of the concentration of the wind power resource in the Atlantic, the North Sea and the Baltic, there has been a suggestion that such a grid would be highly beneficial (see, for example, Reference [3]). It would range from Scandinavia down to France and would link Germany, the UK, Ireland and Spain on the way, as shown in Figure 8.3. The cost of such an enterprise would be substantial but it has been calculated that it would be more cost-effective than the upgrading of the existing onshore networks necessary to absorb the large available offshore potential.

Another advantage would be the considerable aggregation of the resource over a very large geographical area, resulting in a much firmer resource in spite of local variations. Additionally the system would have the ability to provide energy to the country with the greatest need at specific times and the possibility of using the grid as an interconnector between national markets to enhance energy trading. This is particularly attractive because of the time

---

[1] Indeed, a number of these requirements are incorporated into recent Grid Codes, for example those of E.ON Netz, Germany (Grid Code High and Extra High Voltage, E.ON Netz GmbH Bayreuth, August 2003, http://www.eon-netz.com), National Grid, UK (The Grid Code, Issue 3, Revision 19, January 2007) and EirGrid, Ireland (Grid Code V2,0, April 2007).

Figure 8.3 Concept map of the proposed European DC supergrid. (Reproduced with permission of Airtricity)

difference and therefore the time shifts in peak demand among the linked counties. It should be noted that present HVDC systems are usually point to point therefore considerable technology development would be required to deliver a multiterminal HVDC network of the sort proposed.

8.2.3 Building Integrated Wind Turbines

There has been considerable interest of late, particularly in Northern Europe, in the possibility of fitting small wind turbines on the roof tops of individual houses. Enthusiasts for this approach claim that small wind turbines sited within the urban environment have the potential to make a significant contribution to the reduction of CO_2 levels. A typical wind turbine of this type is illustrated in Figure 8.4.

The difficulties facing the designers and manufacturers of such machines are substantial. The dis-economies of downsizing are such that the installation cost per kW of a small wind turbine is considerably more than for a MW size machine. Moreover, the annual average wind speed in a typical urban environment is perhaps one half of that usually experienced in

Figure 8.4 Small building integrated wind turbine. (Reproduced with permission of Windsave)

a good wind farm site. The cubic power law therefore indicates that the expected energy yield per kW capacity from the small wind turbine in these circumstances would be approximately one eighth of that from its large scale counterpart, and this ignores the substantial advantage of the higher wind speeds at the hub height of the large machine. Additionally, in urban locations wind turbulence is considerable due to the presence of buildings and trees and this can affect energy capture and the life of the turbine. On the positive side, small wind turbines if manufactured in, say, hundreds of thousands, can gain enormously from mass production economies of scale and present costs may be more than halved. Government subsidies may accelerate their adoption and their progress down the learning curve.

In general, however, because of long current payback periods, the market for such machines is limited with the unfortunate side effect that the low costs that may result from mass production are unachievable at present. There are in addition concerns about potential noise and vibration problems, as well as the challenge of obtaining planning permission,[2] especially if neighbours object to the installation, and as yet untested arrangements for selling any excess energy fed to the grid. However, there will be some urban areas on hilltops and on coastal locations where the wind speed is high enough, when taken together with escalating electricity costs, to justify the investment in this technology.

Small wind turbines are already economically viable in remote locations not serviced by the grid and for the provision of modest amounts of electricity to caravans and sailing boats. This is an already established market, but one for which the costs of alternative electricity supply is substantially higher than that supplied from the grid.

Enterprising architects have produced futuristic designs that incorporate wind turbines within the building structure, frequently with a ducting effect to enhance the energy capture. One such design is shown in Figure 8.5, where two 43-floor office towers are connected by

---

[2]Interestingly, the UK Government has been considering legislation that could remove the requirement for planning consent for suitable small turbines.

Figure 8.5   The Bahrain World Trade Centre. (Reproduced with permission of the Bahrain World Trade Center image©Khalid Al Muharrqi based on a design by Atkins)

three turbine bridges. Each turbine is 29 metres in diameter and has a rating of 225 kW. The turbines were intended to supplement the energy requirements of this prestigious Bahrain World Trade Centre.

8.3 The Future of Solar Power

As described in Chapter 2, solar technology is developing fast and has much to offer despite the currently high costs. However, to make a significant contribution to bulk electricity generation, a major technology development and cost reduction is required. Possible future developments are discussed in more detail below.

8.3.1 PV Technology Development

The existing PV market is dominated by silicon wafer cells. Commercial modules based on this technology are up to 20% efficient but costly to manufacture and have a high embodied energy content. The energy payback is in years, with the exact figure dependent on the local radiation resource, and contrasts with wind energy where the energy payback period is typically measured in months (see for example, References [4] and [5]).

Thin film PV has the potential to deliver cheaper electricity. Amorphous silicon was the first thin film to become commercially available and in recent years has managed a steady but limited market share [6]. Unfortunately, intrinsic materials issues have limited its performance and it is not expected to be competitive in the longer term. Currently, commercial interest centres around thin film nanocrystalline hybrid cells and the heterojunction cells based on copper indium diselenide and cadmium telluride. Both of the latter provide reasonably stable efficiencies and are expected to benefit from improved manufacturing techniques. There is currently a flurry of investment activity in these technologies that includes manufacturing on to glass, and metal and polymer foils.

Industry experts confidently expect that the new mass production technologies will deliver cells capable of generating electricity competitively with conventional forms and nuclear perhaps within the next 10 years.

Different Deployment Options

The bulk of current grid-connected PV is located on domestic building roofs. Japan and Germany between them have installed over a million such roof systems, primarily as a result of generous market support mechanisms. A similar programme is now planned in the USA. Earlier installations in the main made use of traditional looking modules, although there is now increasing use of so-called PV roof tiles. These can be laid more or less in the same way as traditional tiles and lend the roof a more traditional appearance. Alongside domestic roofs, commercial buildings also provide attractive locations for PV where flat roofs and facades lend themselves to PV installation. There are numerous examples of architecturally appealing PV facades to be found around the cities of Europe. There are thermal issues associated with such building integration of PV. If not carefully designed, the mounting systems can inhibit natural ventilation of the modules, leading to overheating with possible damage to the PV cells in conditions of high radiation and ambient temperature. Fortunately these issues are now fairly well understood.

In contrast to the various forms of building integrated PV, large scale arrays of PV modules can be located on dedicated, usually rural, sites. Systems of up to 10 MW are being installed in the sunnier parts of Europe. A computer image of the world's biggest solar power plant under construction in Walpolenz, Germany, is shown in Figure 8.6. Once fully operational Walpolenz will occupy about 110 hectares, will be rated at 40 MW and produce 40 GWh, approximately the annual consumption of 10 000 households.

8.3.2 Solar Thermal Electric Systems

As outlined in Chapter 2, considerable effort has gone into the development of solar thermal electricity generating systems. These are only feasible in large scale developments and as a result the present solar thermal installations are larger than the PV based ones. Figure 8.7 shows the 150 MW parabolic trough plant at Kramer Junction, California. It has been calculated that if an area of hot desert of about 250×250 km – less than 1% of the total areas of the world's deserts – were to be covered by such thermal plant enough energy would be produced to cover all present electricity needs. Manufacturers of such

Figure 8.6 A computer image of the 40 MW Walpolenz PV installation. (Reproduced with permission of the juwi group)

Figure 8.7 The 150 MW solar thermal plant at Kramer Junction, USA. (Source: Solel)

systems are upbeat about the future but only time will tell whether their enthusiasm is justified.

8.4 The Future of Biofuels

Biomass has a special role to play in the sustainable electricity supply systems of the future due to the fact that it is easily stored. Electricity generation based on biofuels, supported by adequate fuel distribution and fuel stores local to the generation plant, is fully dispatchable and thus can be used to compensate for the variations in the nondispatchable renewables and the system load. Whether there will be sufficient biofuels for this purpose will depend on the available land area for energy crop production in relation to the amount of balancing plant required.

Biomass resources in most countries are large; for example in the UK is has been estimated that up to 20 million tonnes could be available per annum. It is important to distinguish between first generation crops that have been developed for food (sugar beet, oil seed rape and wheat grain) that may be used for chemical conversion to biodiesel and bioethanol and second generation lignocellulosic (biomass) crops that can be used as feedstock for heat, power and liquid fuels.

First-generation technologies have in general a poor carbon footprint and represent an 'intermediate step' towards second-generation lignocellulosic feedstock. Research is needed on the environmental impact of the wide-scale growing of biomass crops. At present there is limited understanding on how the different bioenergy chains compare in environmental impact.

There are also concerns that dedicating significant land area to bioenergy crops will put pressure on food production and result in unwanted increases in food costs. Impacts on food prices in the USA and China have been reported recently, but these reflect the use of corn for the production of liquid fuels for transport, rather that fuels for electricity generation.

8.5 The Future of Hydro and Marine Power

Although small scale hydro installations can make a useful local contribution to electricity generation there are a limited number of suitable sites and thus the overall impact on electricity system operation is limited. This is unlikely to change with time. Larger scale hydro is a different matter. In countries like Switzerland and Norway with a large proportion of hydro power, the power systems are already designed around these resources and their characteristics. Some parts of the world have yet to develop their hydro potential fully and as they do so attention will be paid to the creation of a suitable system infrastructure.

Tidal power, as summarized in Chapter 2, has a large contribution to make in geographical areas that experience a large tidal range. To date this resource has barely been touched. Since the technology for barrage construction and turbine construction is already available, it can be concluded that it is unfavourable economics that presently inhibits the use of this technology. It is known that large tidal projects could take decades to pay back their installation costs. Even though the civil works would last well beyond such timescales, nearer term market pressures make these projects impossible to finance. Nevertheless, there will come a time when fossil fuel shortages and rising energy costs will make the construction of large barrage schemes attractive despite the adverse local environmental impact on wading birds and other wildlife. Indeed the recent concerns for security of supply that are giving hope to the nuclear industry should be prompting a re-examination of this technology, which poses investment challenges that are not dissimilar. Both nuclear and hydro involve massive upfront costs and these need to be paid back with profit over their considerable lifetimes. As with nuclear, states will need to decide how to facilitate the construction of large tidal barrage schemes if the conclusion is that they have an important role to play in the long term future electricity supply.

It may be that concerns about the ecological side-effects will result in alternatives to the straightforward barrage schemes. One such alternative has been suggested by FoE in which tidal lagoons are created. These are claimed to be capable of producing even more energy than barrage schemes as well as being less environmentally intrusive. The newer technology of tidal stream, introduced in Chapter 2, suites different sites and is generally seen as promising despite its present lack of development. It is widely anticipated that reliable and cost-effective technology will be developed. This will make a useful contribution, but only in very specific geographical areas where suitable stream velocities exist.

Wave power is perhaps the least developed renewable energy source, although the resource, especially around the UK coastline, is huge. Since 2000, a number of large scale wave power prototypes have been demonstrated around the world, but the technology is still up to 15 years behind wind energy. In the UK only one wave energy device (Pelamis, see Chapter 2) has been demonstrated at near full scale in the open sea. The first commercial wave energy farms using the Pelamis device are being planned in Portugal, Orkney and Cornwall. Designing devices to survive in the open seas is challenging but enthusiasts for the technology are projecting significant electricity generation from this source in the future.

8.6 Distributed Generation and the Shape of Future Networks

Many reports have been published covering the need for changes in the structure of power systems. Key reports include References [7] to [10] and these have been drawn on in this section.

8.6.1 Distribution Network Evolution

Traditionally, engineers have regarded electricity distribution networks primarily as passive conduits of power from the high voltage transmission system to the loads and have been installed on a 'fit-and-forget' basis. However, some active control elements are now being introduced as far down as the 11 kV network for fault restoration to improve customer supply reliability. Some limited voltage control is also incorporated as described in Chapter 6, but these networks can generally been viewed as passive in operation. All this is changing with the growth of distributed generation.

The connection of smaller CCGTs at distribution voltages has been increasingly common in the European community over the last 15 years and similar trends are occurring elsewhere. Much of the UK's wind energy capacity, especially the older smaller wind farms, were connected at 11 kV, and so embedded in the distribution system. In the USA, distributed generation consists primarily of standby diesel generator sets that add to system security by operating at times of high network stress. In addition there is coming on to the market a wide range of very small generators (micro-chp, PV and micro wind) that are designed for use in domestic and small commercial premises. These trends are expected to continue. As regional penetrations rise new ways of designing and operating distribution networks will evolve. There are presently a number of major research programmes around the world specifically charged with investigating these future scenarios and coming up with technological solutions to the operational challenges that can in the end deliver reliable power supplies at reduced cost.

Changes to be expected will include active and automatic network reconfiguration schemes aimed at facilitating a high penetration of distributed generation while reducing customer minutes lost due to faults.

8.6.2 Active Networks

The network above the 11 kV system is actively controlled with activity increasing as the voltage is increased. For example, the 30/10 kV tap-change transformer in Figure 5.11 is fitted with an automatic voltage controller that monitors the low voltage level and adjusts the taps to keep it within specified limits. Higher up in the chain, at the transmission level, not only active but also reactive power flows are appropriately regulated down transmission lines to ensure that the voltage profile is acceptable, the system is secure under contingencies and that it operates as economically as possible. Additionally under fault conditions, reclose circuit breakers are controlled to improve reliability of supply.

Active control involves some or all of the following functions performed on a continuous basis:

- the measurement of system parameters, for example V, I, P, Q, f and network topology;
- comparing these measurements with set values or with desirable values obtained from analytical computations of the network model, for example load flow analysis, contingency analysis, optimum economic dispatch etc;
- activating a control element to implement the desirable changes, for example increasing the excitation of or the steam input to a synchronous generator, opening or closing a transmission line;
- checking that the changes have been implemented and warning control engineers if the control actions fail.

Distribution networks are extremely extensive. For example, the UK has over 290 000 km of overhead and 470 000 km of underground cable. The conversion of this mostly 'passive' network into an 'active' network would be a major task and will take place gradually as distributed generation is adopted. A key fact is that distribution networks in developed counties have been in place for several decades and will require updating soon due to ageing. This is a unique opportunity to enforce a culture change and design in the new networks the capability of embracing embedded generation rather than repeat yesterday's technology.

In the UK, Ofgem, the Office of Gas and Electricity Markets in a recent consultation document has established an incentive mechanism to encourage Distribution Network Operators (DNOs) to establish what are known as Registered Power Zones (RPZs). These are 'nursery sites' in which DNOs can develop and demonstrate cost-effective ways to connect distributed generation to networks. Typical projects within the RPZs focus on active voltage control, fault level management and power flow control.

8.6.3 Problems Associated with Distributed Generation

The introduction of embedded generation into distribution networks will have a number of technical implications in the following areas.

Fault Levels

In urban areas the distribution networks are designed to have as high a short-circuit level as possible. This is helpful in minimizing the effect that one consumer has on another and in maintaining the consumer voltage as close as possible to the nominal level. Economics dictate that distribution transformers, circuit breakers and cables are rated as near to their maximum load as possible. This small margin between operation and rating implies that if embedded generation is installed, the short circuit level may increase beyond the plant capabilities.

Voltage Levels

One feature of distribution of radial circuits is that they supply a number of distributed consumers; therefore economics dictate that they taper along their length. Embedded generation connected to the end of a long rural circuit is likely to increase the local voltage above statutory levels.

Network Security

Connection of embedded generation has to comply with planning standards that aim to maintain the supply security at the pre-connection level. This may adversely limit the size and type of the embedded generator. The possibility exists that under fault conditions where the supply from the grid is lost, the local system is run in island mode fed solely from the embedded generator(s). In this case, embedded generation will enhance security. This mode of operation requires local active control and is addressed later.

Network Stability

Under certain fault conditions, the system dynamics are excited and it is possible that the characteristics of an embedded generator are such that the ensuing oscillations may cause tripping of the local network. If the dynamics of the generator are known, a stability study can be carried out before connection and if instabilities are detected, control systems theory can devise stabilizing networks.

8.6.4 Options to Resolve Technical Difficulties

Planning Standards

Present planning standards are largely deterministic and therefore conservative. New standards could be developed that rely on the probabilistic nature of renewable energy and of demand patterns. Such standards would be more forgiving and reduce the connection costs and penalties imposed on embedded generation.

Using Power Electronics Technology

High power electronics have made enormous advances over the last decade. These advances have resulted in new devices capable of increasing the flexibility of AC power systems. Such devices, known collectively as flexible AC transmission systems (FACTS), offer considerable opportunities and challenges to power systems engineers. The challenges stem from the necessity of extending as far as possible the capacity of existing transmission lines without affecting adversely the voltage profile and jeopardizing stability. Building additional trans- mission facilities would have solved some of the problems but these are often limited by economic and environmental considerations. FACTS covers a range of electronic based systems such as Static Var compensators, thyristor controlled capacitors, phase-shift trans- formers, interphase power controllers and unified power flow controllers. Some of these devices have been used to assist in the integration of RE sources and are likely to facilitate future connections.

The nature of RE sources are such that they often require a power electronic interface, usually a PWM converter, for connection to the grid. This interface could be regulated, not only to utilize optimally the renewable energy resource, but also to make the generator 'con- nection friendly'. The local or some other vital neighbouring voltage can be monitored and regulated by reactive power injection/extraction. The active power injected into the node may be limited at times to safeguard other local operating restrictions or requirements. The inter- face could be made sensitive to local faults so that it limits the contribution made by the local generator but does not limit it to the extent of endangering network stability. Finally, it could sense the loss of mains and arrange that the local network operates as an island arranging synchronization and reconnection when the mains is re-established.

Such a powerful local control activity will require substantial information processing and transfer. Interactive local control systems could be developed so that the embedded generator interface is in communication with other similar local units such as FACTS and with the local tap-change transformers, so that the most appropriate control actions are taken. The required communications may be provided through the Internet.

Islanding

An actively managed distribution network with locally generated power that approximately balances the demand could theoretically be run autonomously if connection to the mains were to be lost due to a fault. Such operation would require some sophisticated local control actions to first maintain the frequency close to the nominal value and then to reconnect the islanded system back to the mains when the fault is cleared. The capability to island could increase the reliability of supply to the local consumers but at the cost of more complex software and hardware. The desirability of such operation depends crucially on the typical reliability of supply experienced by the consumers. If the reliability is high, say one short interruption per year on average, the benefits of islanding may be doubtful.

Dynamic Loads

In Section 3.8.3 an application of a dynamic load was discussed as a means of regulating frequency, i.e. providing power balance in a power system. In such an application the first requirement is to be able to measure the power system's frequency accurately. A controller needs a cost-effective way to measure the system frequency by, say, using a simple micro-controller to count the number of processor cycles in a full AC cycle. The controller then needs to be aware of the needs of the appliance. In the case of a refrigerator, this means simply measuring the internal air temperature. A relay or other switching device is then added so the controller can switch the compressor (in the case of a refrigerator) or the heating element in the case of an immersion heater.

This application of dynamic load control addresses the totality of system power balance. It is, of course, possible also to use this technology to service local distribution network requirements. For example, the totality of the dynamic loads at the end of a feeder supplying a large group of houses may be diverted from system frequency control to local needs, for example matching from instant to instant the output variations of local embedded generation or assisting the maintenance of system security. Demand-side management is a topic that has been discussed for a long time in power system circles but its considerable potential has not yet been exploited. A major rethink is required here.

Storage

There is at present considerable interest in the possible applications of energy storage in power systems. The technologies are diverse and at different stages of development. They range from a variety of batteries, high speed flywheels, supercapacitors and regenerative fuel cells. Local energy storage would assist embedded generation from renewable energy by providing a buffer between the variability of supply and demand. As the energy storage device would be more than likely to be interfaced to the mains through a power electronic converter, the capabilities of such a converter would also be available to assist the local network.

At the limit, the extensive uptake of embedded generation in the domestic area, such as PV, fuel cells and microturbines combined with improved efficiency of domestic appliances and levels of insulation, could make homes self-sufficient in terms of energy. If such technologies were combined with energy storage, the nature of the distribution network may decline into a minimum infrastructure system providing only occasional energy delivery. The idea of energy self-sufficient homes where the peak to average electricity demand ratio is as high as 10 to 20 (Figure 3.1) is unlikely to be cost effective compared to the sub-stantial improvement that can be achieved in that ratio by grouping several houses together (Figure 3.2). Such small groupings of consumers that benefit from the averaging effect of aggregation is a more likely candidate for future development and it is referred to as a 'microgrid'.

Microgrids

Microgrids may consist of a small community, a housing estate, a university, a commercial or industrial area or a municipality. The key concept is that it includes a number of consumers and a number of small generators (often referred to as microgenerators), located in close proximity so that transmission losses are minimized. The totality of the installed capacity in a microgrid is capable of supplying the local demand. Microgrids promise substantial environmental benefits by facilitating the integration of renewables such as PV arrays and small wind turbines. The microgrid principle is heavily reliant on micro-CHP generators that are driven by the production of heat mostly during winter and PV generators that provide electrical energy during summer. Some storage capability is also assumed. The correct mix is obviously important but would be facilitated by the diversity of the consumer group.

Virtual Power Stations

Virtual power stations are a new concept. A virtual power plant is defined by Wikipedia as a cluster of distributed generation installations (such as micro-CHP, wind turbines, small hydro, back-up gensets, etc.) which are collectively run by a central control entity. Concerted operation of this plant can in theory provide network services such as power balancing and frequency control. The concept is a response to market mechanisms that penalize the variability and lack of predictability of small individual generators, particularly those based on renewable energy sources. However, since these aggregations are expected to reflect commercial arrangements that might involve highly dispersed plant, rather than geographically specific clusters, the operation of such virtual plants might conflict with the requirements for local power and voltage control. These issues are the topic of ongoing research and so far there is no clear consensus on the value of this approach.

8.7 Conclusions [11, 12]

The nationwide and local electricity grids, metering systems and regulatory arrangements that were created for a world of large-scale, centralised power stations will need restructuring over the next 20 years to support the emergence of far more renewables and small-scale, distributed electricity generation [11].

Intermittency is not a 'flaw' or shortcoming as traditional 'reliability' concepts imply. On the contrary, requiring a system to always deliver generation that matches a fleeting peak load gives rise to a set of generation and network assets that are invariably drastically overspecified and underemployed, a situation long overdue for frontal attack by innovative policy [12].

When a new process technology – wind – is 'shoe-horned' into an existing system that evolved to support previous vintage technology, things do not work correctly. To the network operator it may seem as plain as day that the new technology needs to morph so that it take on the characteristics resembling the old technology. This is wasteful and foolish. Quite the opposite must happen. If we were to effectively exploit technological progress, it is the underlying system that needs to metamorphose and adjust to the innovation [12].

Compared to the breathtaking speed of developments in IT, the technology of power generation and distribution is depressingly lethargic. One reason is that the natural lifecycle of asset renewal in power systems is measured in decades rather than years. In truth, the vast range of IT technology has barely affected the electricity networks. Interestingly, the long delayed revolution required will be precipitated by the imminent impact of renewable sources. The distributed and variable nature of renewables will require intelligently controlled power electronic hardware to maintain power quality and optimize energy flows. This required revolution in flexibility and controllability of the network will, of course, be of immense benefit to traditional generation as well. Much of the distribution network in most countries is at least 40 years old and will have to be replaced over the next decade. There is therefore a great opportunity to introduce innovative technologies rather than replace like with like.

Traditionally, power systems have been run on the policy that whenever the consumer demands energy this should be instantly satisfied from dispatchable generation. With increasing proportions of renewable generation this philosophy must be changed but without substantial inconvenience to the consumer. Demand-side flexibility will require that consumers adjust their demand profile to meet supply through deferrable loads. This will require intelligent control and could be done through time-of-use tariffs requiring customer voluntary response, to hard wired direct control and everything in between. If demand can shift timewise to suit generation, the availability of primary energy will be dictating the pattern of consumption rather than the opposite. In such a power system 'base load' electricity may not continue to have the significance it has at present.

An active distribution network is by definition more complex than the current system, and building and managing it will require new skills. In this future power system, renewable generation will be injected and active control will be implemented at all system levels.

To conclude, a fundamental rethink will be needed of the way electricity is generated, transmitted and distributed in the power systems of the future. Variable generation infeeds coupled with bidirectional network flows will become commonplace. This rethink does not only require the adoption of new technologies but also a change in the mindset or 'culture' of power system engineers. This book has endeavoured to cover the basic principles of how power systems will evolve from the present time when renewable energy contributes a minute proportion towards electricity production to a future when this proportion becomes substantial.

References

[1] A readable overview of HVDC technology can be found in the ABB site www.abb.com/hvdc.
[2] Jacobson, B. et al. 'HVDC with voltage source converters and extruded cables for up to ±300 kV and 1000 MW', CIGRE Paper B4-105, 2006.
[3] O'Connor, E. 'European unity, a vision for sustainable power in Europe', *Renewable Energy World*, March–April 2006, 124–127.
[4] Danish Wind Energy Association website www.windpower.org.
[5] Gagnon, L., Berlanger, C. and Uchiyama, Y. 'Life-cycle assessment of electricity generation options', *Energy Policy*, 2002, 30, 14.
[6] van Sark, W.G.J.H.M., Brandsen, G.W., Fleuster, M. and Hekkert, M.P. 'Analysis of the silicon market: will thin films profit?' *Energy Policy*, June 2007, 35 (6), 3121–3125.
[7] 'Decentralising power: an energy revolution for the 21st century', Greenpeace Publication, 2006.
[8] 'Planning of the grid integration of wind energy in Germany onshore and offshore up to 2020', DENA Study, 2005.

[9] Markvart, T. and Arnold, R. 'Microgrids: power systems for the 21st century', Ingenia, http://eprints.soton. ac.uk/23738/.

[10] 'Network management systems for active distribution networks', DTI Study No. KEL00310, 2004.

[11] 'Our common future', Energy White Paper, DTI, UK, p. 16.

[12] Awerbuch, C. 'Decentralisation, mass-customisation and intermittent renewables in the 21st century: restructuring our electricity networks to promote decarbonisation', Tyndall Working Paper Series, March 2004, pp.14 and 13.

Appendix: Basic Electric Power Engineering Concepts

A.1 Introduction

This appendix covers some of the very basic topics that underlie the generation, transmission and use of electrical energy, but it is not intended to be a substitute for a good textbook on basic electrical engineering. For readers who have no background whatsoever in this topic a selective study of relevant chapters in a typical good textbook, as, for example, Reference [1], would be essential.

This appendix is intended for readers who have studied 'electricity' as a subsidiary subject at university and need to refresh their memories. It is likely that the 'electricity' they have met was slanted towards the understanding of electronic circuits and devices. This is because the major concern of electrical engineering since the invention of the transistor and of the computer has been the 'communications' and 'control' rather than the 'energy' aspects of the discipline. It is therefore possible that, although such readers are knowledgeable in some aspects of electricity, they are unfamiliar with the ideas and conventions developed over the years by power system engineers. If this is the case, they are likely to find this appendix useful.

For a power system to be an efficient channel of energy flow from generators to consumers the task should be accomplished with the minimum energy loss. Because of the need to supply a constant voltage, it is also necessary that the desirable energy transfers are achieved within specified upper and lower voltage limits at the system junctions or nodes. These requirements have led to certain analytical concepts that are particular to power systems engineering.

A.2 Generators and Consumers of Energy

A start is made with the definition of the conventions used when power is to be calculated in electric circuits. In Figure A.1 the box represents any single piece or collection of

Renewable Energy in Power Systems   Leon Freris and David Infield
© 2008 John Wiley & Sons, Ltd

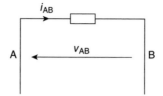

Figure A.1   Circuit element

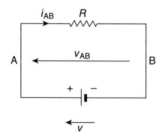

Figure A.2   DC circuit example

equipment found in a power systems network. In the analysis that follows, whenever lower case letters are used for circuit variables such as voltage, current and power, it will be taken that these refer to instantaneous values. The voltage v_{AB} describes the potential difference of terminal A with respect to terminal B at one instant. If the potential of A is higher than that of B then v_{AB} is positive. It follows that $v_{BA} = -v_{AB}$. Now i_{AB} will be defined as the current that flows from A to B *through* the element. If the current transfers positive charge from A to B then i_{AB} is said to be positive. Again it follows that $i_{BA} = -i_{AB}$.

Power in electricity is given by the product of voltage and current. For the element in Figure A.1 the instantaneous power p is given by

$$p = v_{AB}i_{AB} \tag{A.1}$$

Figure A.1 encapsulates the convention for the reference voltage and current directions in this simple circuit configuration which, for obvious reasons, is known as a *one-port*. When setting up the reference directions in a one-port it is said that one looks into the terminals AB towards the circuit in question.

If v_{AB} and i_{AB} are both positive then p is also positive and the element is a *consumer* or *sink* of electrical power. The truth of this simple convention can be tested by examining the simple DC circuit of Figure A.2. Here the circuit element is a plain resistor and terminals AB are connected to a battery of the polarity shown. With this DC *excitation* the instantaneous quantities are of course the same as the steady state values; v_{AB} and i_{AB} are both positive in this case and the power p associated with the resistor is consumed and irreversibly converted into heat. Figure A.3 is identical to Figure A.2 but now the battery can be considered as being the element in the box of Figure A.1. Therefore the terminals AB are looked into downwards rather than upwards, with the result that current i_{AB} has reversed direction.

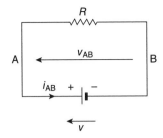

Figure A.3   DC circuit example

Current i_{AB} through the battery is now negative, v_{AB} is still positive and therefore Equation (A.1) gives a negative value to the power associated with the battery. If a line is drawn at the interface AB, power is going into the interface if looked at towards the resistor, i.e. the power is positive, but power is coming out of the interface if looked at towards the battery, i.e. the power is negative. This, of course, satisfies the conservation of energy principle at the interface.

To sum up, the power associated with an element is taken as positive and therefore consumed when the current is flowing *through* the element from higher to lower potentials. If the reverse were the case, the power would be negative and the element would be a generator. All this is self-evident when dealing with DC circuits, but what would be the power associated with an element if v_{AB} and i_{AB} were continuously varying quantities such as sinusoids?

A.3 Why AC?

In some cases AC holds no practical advantage over DC. In applications where electricity is used to dissipate energy in the form of heat, the polarity or direction of current is irrelevant. However, with AC it is possible to build electric generators, motors and power distribution systems that are far more efficient or flexible than with DC, and so AC is used predominately across the world in high power applications. Section 4.2 provides information on how AC is generated through a rotating mechanical–electrical energy converter.

The relative simplicity of AC generators and motors translates into greater reliability and lower cost of manufacture. There is, however, an additional very important advantage of AC over DC. If there are two mutually inductive coils and one coil is energized with AC, an AC voltage will be created in the adjacent coil. This device is known as a transformer and its mode of operation is described in some detail in Section 4.3. The fundamental significance of a transformer is its ability to step voltage up or down from the powered coil to the unpowered coil. As explained in Section 1.4 of Chapter 1, the transformer's ability gives AC an advantage unmatched by DC in the realm of power transmission and distribution.

A.4 AC Waveforms

In practice AC generators are designed to produce a voltage that, to all intents and purposes, is a pure *sine wave*. There is a good reason why pure sign waves are used in power networks.

A sine wave is the only waveform that, when differentiated or integrated, results in a replica of itself. As will be shown later, electric power system components, including transformers, respond differentially to a sine wave excitation and thus the shape of the waveform is maintained throughout the network. Additionally, three-phase electric motors are at their most efficient and produce a constant torque when fed from an AC supply of sine wave shape only. A sinusoidal voltage of frequency f can be written analytically as $v = \hat{V} \sin \omega t$, where $\hat{V}$ is the peak value of the sine wave and $\omega = 2\pi f$.

From now onwards, whenever AC is referred to, it is implied that the waveform is of sinusoidal shape. With AC a measurement dilemma is encountered in expressing how large or small an AC quantity is. With DC, where quantities of voltage and current are generally stable, there is no trouble expressing what is the value of the voltage or current in any part of a circuit.

One way to express the magnitude (or *amplitude*) of an AC quantity is to measure its peak value $\hat{V}$. However, when dealing with measurements of electric power, the best way to express the value of an AC waveform is to label it in terms of its ability to perform useful work. The way to do this is to assign to the waveform a 'DC equivalent' measurement, i.e. to label it by the value in volts or amperes of the DC that would have produced the same amount of heat energy dissipation in the same resistance. The mathematical method of producing the DC equivalent value, known as root mean square (RMS), can be found in any elementary textbook on electricity. In this book the RMS value of an AC voltage or current will be denoted by an italic capital letter, for example V, I. It can be shown that for a sine wave the RMS value is related to the peak by

$$V = \sqrt{2}\hat{V} = 0.707\hat{V} \qquad (A.2)$$

Thus the household voltage of 230 V has a peak of 325 V.

A.5 Response of Circuit Components to AC

Power networks consist mainly of generators, transmission lines, transformers and consumers. Each one of these constituent parts consists in turn of a combination of three basic components, namely resistance R, inductance L and capacitance C. Additionally, generators have internally induced AC voltages caused by the interaction of moving conductors and magnetic fields. These electromagnetic actions are the mechanisms by which mechanical energy, conventional or renewable, is converted into electrical form. Uniquely, in photovoltaic systems the conversion of solar power is carried out directly into electricity without the aid of electromagnetics.

The AC source voltages are the drivers of all current flows in the network and therefore partly determine the distribution of power flows from the generators to consumers over the transmission network. The energy flows in a network consisting of traditional and/or renewable generators can be determined by the application of Kirchhoff's laws, but for this to be done the 'resistance' to the flow of AC current offered by the three basic components must be known. In other words, the relationship between AC voltage and current (Ohm's law) for each component must be determined.

A.5.1 Resistance

Ohm's Law for a resistance R in ohms is the linear relationship

$$i = \frac{v}{R} \tag{A.3}$$

where v in volts and i in amperes are the instantaneous values of voltage and current. Figure A.4 shows a plot of the applied voltage

$$v = \hat{V} \sin \omega t \tag{A.4}$$

and the resulting current i in a circuit consisting of an AC source and a resistor R. From Equation (A.3) the resulting current is given by

$$i = \frac{\hat{V}}{R} \sin \omega t \tag{A.5}$$

Equation (A.1) shows that the instantaneous power is given by

$$p = vi = \frac{\hat{V}^2}{R} \sin^2 \omega t = 2 \frac{V^2}{R} \sin^2 \omega t = \frac{V^2}{R}(1 - \cos 2\omega t) \tag{A.6}$$

where V is the RMS value of the AC voltage.

Equation (A.6) is plotted in Figure A.5. This is a double frequency waveform of average value V^2/R and is fully displaced above the time axis. This confirms the evident fact that the power converted in the resistor from electrical into heat form is not dependent on the

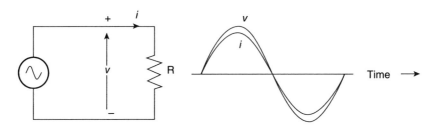

Figure A.4   Resistor excited by AC voltage

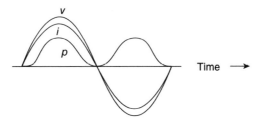

Figure A.5   Power in a resistive component

direction of current. It is concluded that to calculate the power in a purely resistive element any of the following expressions can be used:

$$P = \frac{V^2}{R} = RI^2 = VI \qquad (A.7)$$

A.5.2 Inductance

Inductors do not behave like resistors. Whereas resistors simply oppose the flow of electrons through them (by dropping a voltage directly proportional to the current), inductors oppose *changes* in current through them, by dropping a voltage directly proportional to the *rate of change* of the current.

Ohm's law for an inductance L in henries is the following differential rather than linear relationship:

$$v = L\frac{di}{dt} \qquad (A.8)$$

Assume that a sinusoidal current $i = \hat{I}\sin\omega t$ flows through an inductor of inductance L. Inserting this in the r.h.s. of Equation (A.8) gives $v = (\omega L)\hat{I}\cos\omega t = \hat{V}\cos\omega t$, where

$$\hat{V} = (\omega L)\hat{I} \qquad (A.9)$$

Figure A.6 shows the voltage/current relationship for an inductor. The voltage represented by the cosine curve is $90°$ ahead of the current sine curve. The voltage is now said to *lead* the current by $90°$ or the current to *lag* the voltage by the same amount. In general, if two AC waveforms are phase-shifted by $90°$, it is said that these waveforms are in *quadrature* to each other. If the voltage is described by

$$v = \hat{V}\sin\omega t \qquad (A.10)$$

then the current would take the form $i = \hat{I}\sin(\omega t - \pi/2)$, which through Equation (A.9) becomes

$$i = \frac{\hat{V}}{\omega L}\sin\left(\omega t - \frac{\pi}{2}\right) \qquad (A.11)$$

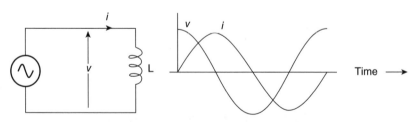

Figure A.6   Inductor excited by AC voltage

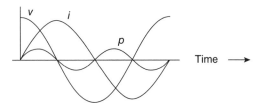

Figure A.7   Power in an inductive component

The instantaneous power is

$$p = vi = \widehat{V}\widehat{I}\cos\omega t\sin\omega t$$

and through Equation (A.9),

$$p = (\omega L)\widehat{I}^2\sin\omega t\cos\omega t = (\omega L)I^2 2\sin\omega t\cos\omega t = (\omega L)I^2\sin 2\omega t \qquad (A.12)$$

where I is the RMS value of the current.

Equation (A.12) is plotted in Figure A.7. Here the power is of double frequency, as in the case of the resistor, but it is wholly symmetrical with respect to the time axis. This means that a perfect inductor (one that possesses no winding resistance) is not associated with a net power transfer over the long run. During periods in the cycle when the power is positive, the inductor absorbs energy from the network to which it is connected and stores it in its magnetic field. During periods when the power is negative, the stored energy is returned to the rest of the system. It can be concluded that an inductor acts as a consumer or sink of electrical energy during part of the cycle and a generator of energy during the remaining cycle.

The purpose of power systems is to transfer electrical power from generators to consumers so that the latter could convert this energy usefully into heat, light, mechanical or chemical power or use it to drive IT systems. The power associated with an inductor serves none of these purposes. It consists of power oscillating between power system components.

A.5.3 Capacitance

Capacitors oppose changes in voltage by drawing or supplying current as they charge or discharge to the new voltage level (Figure A.8). The flow of electrons through a capacitor is directly proportional to the rate of change of voltage across the capacitor. This opposition to voltage change is a mirror image to the kind exhibited by inductors.

Ohm's law for a capacitance C in farads is the differential relationship

$$i = C\frac{dv}{dt} \qquad (A.13)$$

If a sinusoidal voltage

$$v = \widehat{V}\sin\omega t \qquad (A.14)$$

is applied to a capacitor, the resulting current from Equation (A.13) is

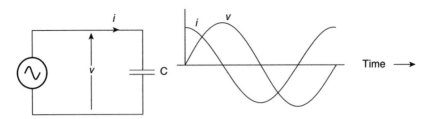

Figure A.8 Capacitor excited by AC voltage

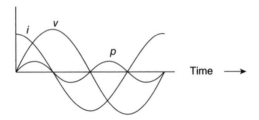

Figure A.9 Power in a capacitive component

$$i = (\omega C)\hat{V}\cos\omega t = \hat{I}\sin\left(\omega t + \frac{\pi}{2}\right) \qquad (A.15)$$

where

$$\hat{I} = (\omega C)\hat{V} \qquad (A.16)$$

The current is now said to *lead* the voltage by 90° or the voltage to *lag* the current by the same amount.

The simple capacitor circuit is characterized by a power curve (Figure A.9), which is similar to that of the pure inductor.

A capacitor, also a storage element, does not dissipate power as it reacts against changes in voltage; it merely absorbs and releases power, alternately.

There is one final observation that must be stressed. A comparison between the sinusoidal power waveforms in Figures A.7 and A.9 reveals that they are always in opposition when viewed with respect to the voltage waveform. That is to say, during periods when the power in the inductance is positive, the power in the capacitance is negative and vice versa. This is an important fact that will be returned to later.

A.6 Phasors

In the above analysis the AC variations of voltage, current and power were depicted as sinusoidally varying functions of time. This is acceptable for an introductory exposition of the relationships of these quantities for single-circuit elements. In real life, power circuits consist

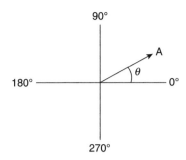

Figure A.10   The Argand diagram of complex numbers

of numerous combinations of such elements and a more approprite tool is required for their analysis.

The kind of information that expresses a single dimension, such as linear distance or temperature, is called a *scalar* quantity. The voltage produced by a battery, for example, is a scalar quantity. However, when alternating current circuits are analysed, it is found that voltage and current are not the familiar one-dimensional quantities. Rather, these quantities, because they are dynamic (alternating in direction and amplitude), possess two dimensions, i.e. amplitude and phase. Therefore there is a need to work with mathematical techniques capable of representing two-dimensional quantities. A *complex number* is a single mathematical quantity able to express these two dimensions of amplitude and phase at the same time. A graphic representation of a complex number is known as a *phasor*.

It must be obvious that there must be some common frame of reference for angles to have any meaning. In this case, directly right in Figure A.10 is considered to be $0°$ and angles are counted in a positive direction going counterclockwise.

The length of the phasor represents the *magnitude* (or amplitude) of the AC waveform. The greater the magnitude of the waveform, the greater the length of its corresponding phasor. The angle of the phasor represents its *phase*, i.e. the phase shift in degrees between the waveform in question and another waveform acting as a 'reference' in time. In this book a phasor is denoted by a bold capital letter, for example *A*. Phasor *A* in Figure A.10 has a magnitude *A* and a phase angle θ. The representation of a phasor by magnitude and phase is known as its *polar* form. The polar form of phasor *A* in Figure A.10 is written as $A = A \angle \theta$. The phase of a waveform in a circuit is usually expressed with regard to the power supply voltage waveform (arbitrarily stated to be 'at' $0°$). If there is more than one AC voltage source, then one of those sources is arbitrarily chosen to be the phase reference for all other measurements in the circuit. Phase is always a *relative* measurement between two waveforms rather than an absolute property.

In Figure A.11, A and B could represent the voltage and current waveforms, (a) for a resistor, (b) for an inductor and (c) for a capacitor; (d) could represent the antiphase relationship of power in an inductor and a capacitor.

A.7 Phasor Addition

If phasors that are not in-phase or antiphase are added, their magnitudes add up quite differently to that of scalar quantities. Figure A.12 shows an example of such an addition.

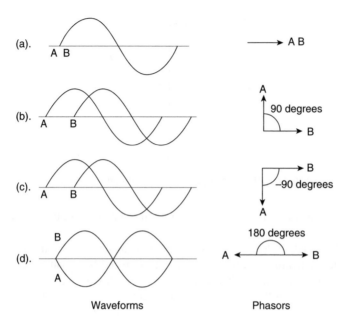

Figure A.11   Phasor representation of AC quantities

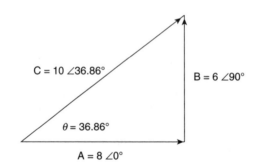

Figure A.12   Addition of phasors

In fact, phasors add up in exactly the same way as vectors that represent velocities or forces in mechanical systems. As explained in Section A.4 the 'effective' value of an AC waveform is its RMS value; hence AC quantities, unless stated otherwise, are always labelled in terms of their RMS value and are expressed by a capital letter. For example, V and I represent the RMS values of the general sinusoids

$$v = \sqrt{2}V \sin(\omega t + \phi) = \hat{V} \sin(\omega t + \phi) \qquad \text{and} \qquad i = \sqrt{2}I \sin(\omega t + \theta) = \hat{I} \sin(\omega t + \theta)$$

respectively, where $\hat{V}$ and $\hat{I}$ are the peak values and ϕ and θ are phase angles with respect to a reference sinusoid.

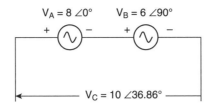

Figure A.13 Addition of AC voltages

If two AC voltages are added together by being connected in series as in Figure A.13, their voltage magnitudes do not directly add or subtract as with scalar voltages in DC. Instead, these voltages are complex quantities, and just like the above phasors, which add up in a trigonometric fashion, an 8 volt source at $0°$ added to a 6 volt source at $90°$ (both RMS) results in 10 volts at a phase angle of $36.86°$.

The property of phasors that, although stationary on a page, enables them to represent AC time varying sinusoidal quantities is of inestimable value in the calculation of power flows in complex AC circuits containing voltage sources and combinations of R, L and C elements.

A.8 Rectangular Notation

Simple electric networks can be analysed through phasor diagrams. This is extremely tedious when a more challenging circuit is involved and becomes impossible for real life networks. Phasors can be represented mathematically by complex numbers and circuits can be analysed through algebraic manipulations without the laborious task of drawing phasors. There are two basic forms of complex number notation: the polar, described earlier, and the *rectangular* or *Cartesian*.

In the rectangular form, a complex number is denoted by its respective horizontal and vertical components. In order to distinguish the horizontal and vertical dimensions from each other, the vertical is prefixed with a lower-case j. The horizontal and vertical phasor components are known as real and imaginary numbers respectively.

Figure A.14 shows the 'complex number plane' on which any phasor expressed in rectangular coordinates can be drawn. The reader must know from his/her courses in mathematics that $j = \sqrt{-1}$ and therefore $j \times j = -1$. The geometric interpretation of this in the complex plane is that multiplication by j rotates a phasor anticlockwise by $90°$.

In Figure A.15 the phasor V of magnitude V and phase angle θ can be written in rectangular or polar form as follows: $V = a + jb = V \angle \theta$. The conversion between rectangular and polar is carried out through

$$V = \sqrt{a^2 + b^2} \qquad \text{and} \qquad \theta = \tan^{-1}\frac{b}{a}$$

To convert from polar to rectangular:

$$a = V\cos\theta \qquad \text{and} \qquad b = V\sin\theta$$

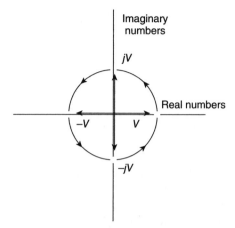

Figure A.14 Complex number plane

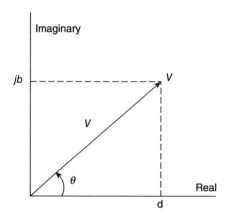

Figure A.15 Rectangular–polar conversion

It is also possible to write $V = V(\cos\theta + j\sin\theta)$. A famous expression known as Euler's theorem states that $\cos\theta + j\sin\theta = e^{j\theta}$; therefore an alternative expression for V in polar form is

$$V = Ve^{j\theta} \tag{A.17}$$

Circuit calculations involving the addition or subtraction of complex numbers are carried out conveniently using rectangular coordinates. Polar coordinates are more convenient when multiplications and divisions are required. Since complex numbers are legitimate mathematical entities, just like scalar numbers they can be added, subtracted, multiplied, divided, squared, inverted and so on, just like any other kind of number.

Addition and subtraction with complex numbers in rectangular form is easy. For addition, the real components are simply added to determine the real component of the sum and the imaginary components are added to determine the imaginary component of the sum. For subtraction, just take the difference of these numbers. When multiplying complex numbers in polar form, simply multiply the polar magnitudes of the complex numbers to determine the polar magnitude of the product and add the angles of the complex numbers to determine the angle of the product. If dividing, just divide the magnitudes and subtract the angles. Meter measurements in an AC circuit correspond to the *polar magnitudes* of calculated values. Rectangular expressions of complex quantities in an AC circuit have no direct, empirical equivalent, although they are convenient for calculations.

All the rules and laws of DC circuits, i.e Ohm's law, Kirchhoff's laws and network analysis methods, apply to AC circuits as well, with the exception of power calculations. The only qualification is that all variables *must* be expressed in complex form, taking into account phase as well as magnitude, and all voltages and currents must be of the same frequency (in order that their phase relationships remain constant).

A.9 Reactance and Impedance

Now expressions will be developed that describe the 'resistance' or 'opposition' offered to the flow of AC currents by the three types of circuit elements. This can be found by dividing the phasor of voltage by the resulting current phasor. The voltage will be taken as the reference phasor and polar notation will be used in these derivations.

A.9.1 Resistance

Dividing Equations (A.4) by (A.5) gives

$$\text{Opposition offered by resistance} = \frac{V\angle 0°}{I\angle 0°} = R \text{ ohm} \tag{A.18}$$

The relationship here is very simple. To derive the current through a resistor just divide the voltage phasor by the resistance. The current and the voltage are in phase.

A.9.2 Inductance

Dividing Equation (A.10) by (A.11),

$$\text{Opposition offered by inductance} = \frac{V\angle 0°}{I\angle -90°} = \frac{V\angle 0°}{V/(\omega L)\angle -90°}$$

$$= \omega L\angle 0° - (-90)° = X_L\angle 90° = jX_L \text{ ohm} \tag{A.19}$$

The quantity $jX_L = j\omega L = j2\pi f L$ is known as *inductive reactance* and represents the 'opposition' to the flow of current when a voltage of frequency f Hz is applied across an inductance of L henry. The 90° phase lag that occurs to the current with respect to the voltage in an inductive element is conveniently incorporated in the description of the reactance through the operator j. There is nothing 'imaginary' of course about a reactor that consists typically of a coil of wire. However, as will be shown later, by assigning the j operator to the reactance the calculations are enormously assisted.

A.9.3 Capacitance

Dividing Equation (A.14) by (A.15),

$$\text{Opposition offered by capacitance} = \frac{V\angle 0°}{I\angle 90°} = \frac{V\angle 0°}{V\omega C\angle 90°}$$

$$= \frac{1}{\omega C}\angle 0° - 90° = X_C \angle -90° = -jX_C \text{ ohm} \qquad (A.20)$$

The quantity $-jX_C = -j/(\omega C) = -j/(2\pi f C)$ is known as *capacitive reactance* and represents the 'opposition' to the flow of current when a voltage of frequency f Hz is applied across a capacitance of C farad. Again the 90° phase lead to the current is conveniently incorporated in the $-j$ operator of the reactance expression.

A.9.4 Impedance

Now two components will be connected in series and the effects will be investigated. Take, for example, the circuit in Figure A.16.

The resistor will offer 5 Ω of resistance to AC current regardless of frequency, while the inductor will offer $2\pi f L = 6.282\,\Omega$ of reactance to AC current at 50 Hz. Because the resistor's resistance is a real number (10 + j0 Ω) and the inductor's reactance is an imaginary number (0 + j6.282 Ω), the combined effect of the two components will be an opposition to current equal to the complex sum of the two numbers. In order to express this opposition succinctly, a more comprehensive term is needed for opposition to the current than either resistance or reactance alone. This term is a complex number known as *impedance* **Z**, and is also given in ohms, just like resistance and reactance. In the above example, the total circuit impedance is

$$\mathbf{Z} = (10 + j0) + (0 + j6.282) = 10 + j6.282 \text{ or } 11.81\angle 32.137°$$

Impedance is related to voltage and current in a manner similar to resistance in Ohm's law:

$$V = IZ \qquad (A.21)$$

where all the quantities are expressed in complex form.

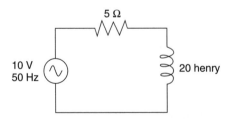

5 Ω

10 V
50 Hz

20 henry

Figure A.16   Series AC circuit

Impedance is a far more comprehensive expression of opposition to the flow of electrons than resistance. Any resistance and any reactance, separately or in combination (series/ parallel), can be represented as a single impedance in an AC circuit.

To calculate current in the above circuit, first the phase angle reference of zero is given to the voltage source. The phase angles of resistive and inductive impedance are *always* $0°$ and $+90°$ respectively, regardless of the given phase angles for voltage or current:

$$I = \frac{V}{Z} = \frac{10\angle 0°}{11.81\angle 32.137} = 0.846\angle -32.137$$

As with the purely inductive circuit, the current sine wave lags behind the voltage sine wave, although this time the angle lag is only $32.137°$.

The same rules applied in the analysis of DC circuits apply to AC circuits as well, with the caveat that all quantities must be represented and calculated in complex rather than scalar form. So long as phase shift is properly represented in the calculations, there is no fundamental difference in how basic AC circuit analysis is approached versus DC.

Now a generalized expression of the complex impedance of a series circuit containing R, L and C will be written:

$$Z = R + jX = R + jX_L - jX_C = R + j(X_L - X_C) = Z\angle\theta \tag{A.22}$$

where

$$Z = \sqrt{R^2 + (X_L - X_C)^2} \quad \text{and} \quad \theta = \tan^{-1}\frac{X_L - X_C}{R}$$

For an applied voltage V, the current in the circuit is given simply by $I = V/Z$. For a circuit containing a number of parallel branches Z_1, Z_2, Z_3 etc., the equivalent impedance is given by

$$Z = \frac{1}{1/Z_1 + 1/Z_2 + 1/Z_3} \tag{A.23}$$

A.10 Power in AC Circuits

The foundations have now been set for the investigation of power flows in power system networks. In Section A.2 the ideas of generators and consumers of energy were investigated. The concepts are very clear when the quantities involved are considered at one instant in time or are of a DC nature. An investigation will be made of what happens when v_{AB} and i_{AB} are sinusoidal quantities in circuits consisting of generators, consumers plus R, L and C elements. There are partial answers to these questions for purely resistive or reactive elements. It stands to reason that in mixed circuits the power is likely be a double-frequency sinusoid partly displaced about the time axis exhibiting both an average value as well as an oscillating component. In what follows, this will be shown analytically.

Figure A.17 shows the waveforms of voltage, current and power associated with a resistive–inductive element. As expected, the current lags the voltage by an angle more than $0°$ and less than $90°$. The power is the instant-by-instant product of the voltage and current and has both positive and negative values. Hence during parts of the cycle the element acts

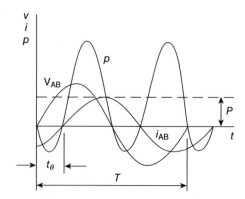

Figure A.17   Power in a mixed circuit

as a consumer and during other parts as a generator of power. What decides whether the element is a consumer or generator of power over the long run is the sign of the integral:

$$P = \frac{1}{T}\int_t^{t+T} p\,dt \tag{A.24}$$

Let $v_{AB} = \hat{V}\sin\omega t$ and $i_{AB} = \hat{I}\sin(\omega t - \theta)$. Then

$$p = v_{AB}i_{AB} = \hat{V}\sin\omega t\,\hat{I}\sin(\omega t - \vartheta)$$

Using a trigonometric identity,

$$p = \frac{1}{2}\hat{V}\,\hat{I}\cos\phi - \frac{1}{2}\hat{V}\,\hat{I}\cos(2\omega t - \theta)$$

Using Equation (A.2) to convert the peak into RMS quantities gives

$$p = VI\cos\phi - VI\cos(2\omega t - \theta)$$

Introducing this expression into the integral of Equation (A.24) and integrating between the limits gives

$$P = VI\cos\theta \tag{A.25}$$

as the second term represents a symmetrical sinusoid, the average value of which over a period T is zero.

Equation (A.25) is of great significance in power systems engineering. The product VI is known as *apparent* power and is measured in *volt-amperes*. The term $\cos\theta$ is known as the *power factor* and θ as the *power factor angle*. It is the product of the power factor with the apparent power that results in the *active power* P which is measured in *watts*. The power factor is unity when the voltage and current are in phase, i.e. for a purely resistive element (Equation (A.7)). These are the only circumstances when the volt-amperes are equal to the watts.

The apparent power, being the product of two RMS scalar quantities that are always positive, is also always positive. It follows that the sign of Equation (A.25), i.e. whether the element is a consumer or a generator of power, depends solely on the sign of $\cos\theta$. It should be clear from Figure A.17 that as the phase-shift angle θ of the current sine wave i_{AB} increases, the average power P will decrease as the power waveform p becomes increasingly symmetrical about the time axis. When θ reaches $90°$, $\cos\theta = 0$ and $P = 0$. For $\theta > 90°$ the average power will reverse sign and the element will generate rather than consume power. It can then be postulated that for $-90° < \theta < 90°$ the element is a consumer of power and that for $90° < \theta < 270°$ the element is a generator of power. When $\theta = 90°$ or $-90°$, the element acts as a pure inductor or capacitor respectively and it is neither generating nor consuming active power.

A.11 Reactive Power

Equation (A.12) gives the instantaneous power for an inductor as $p = (\omega L)I^2 \sin 2\omega t$. It is known that $\omega L = X_L$ and therefore the peak of the power variation is given by

$$I^2 X_L = \frac{V^2}{X_L} = I^2 Z \sin\theta = I^2 \frac{V}{I}\sin\theta = Q = VI\sin\theta \qquad (A.26)$$

The quantity Q is known as reactive power because of its similarity to the active power $P = I^2 R$. It is measured in *reactive volt-amperes* denoted as VAR, pronounced 'vars'. The reactive power Q is of considerable importance in power systems. Notwithstanding the fact that it has zero average value it represents real reciprocating energy transfers between storage elements that are unavoidably parts of power system networks. These transfers are important because they result in extra energy loss in transmission lines (affecting efficiency) and in network voltage changes (affecting adversely, at times, the voltage at consumer terminals).

As with Equation (A.25) for active power, the sign of $\sin\theta$ in Equation (A.26) indicates whether Q is positive or negative. For $0° < \theta < 180°$ the current lags the voltage, $\sin\theta$ is positive and, by analogy to the active power notation, the element is said to be a consumer or a sink of reactive power. For $0° > \theta > 180°$, the current leads the voltage, $\sin\theta$ is negative and the element is said to be a generator of Q. For a capacitive reactance,

$$Q = -I^2 X_C = -\frac{V^2}{X_C}$$

The whole concept is illustrated diagrammatically in Figure A.18, which is similar to the complex number plane of Figure A.14. With the current as the reference phasor the position of V on the diagram defines the nature of a power system element in terms of P and Q. The first quadrant represents elements that are consumers of P and Q, i.e. consist of resistive–inductive components, while the fourth quadrant is for resistive–capacitive elements. Quadrants 2 and 3 require generation of active power and are reserved for 'active' elements such as AC generators. In Chapter 4 it was shown that such a device, known as a synchronous machine, is capable of operating in any of the four quadrants.

A.12 Complex Power

The V and I phasors in Figure A.18 are shown again in Figure A.19(b) in the form of a triangle. With the current as reference, the voltage V applied across the circuit is equal to the

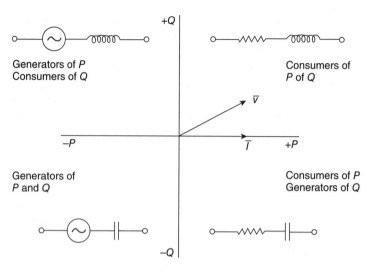

Figure A.18   Quadrant diagram

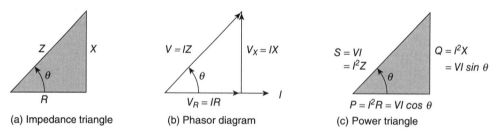

(a) Impedance triangle   (b) Phasor diagram   (c) Power triangle

Figure A.19   The complex power triangle

phasorial addition of the voltage V_R across the resistor (in phase with the current) and voltage V_X across the reactor (in quadrature and leading the current). If the magnitude of each voltage phasor is divided by I a similar triangle, the *impedance triangle*, is obtained, shown in Figure A.19(a). If each voltage magnitude is multiplied by I, another similar triangle, the *power triangle*, is obtained, shown in Figure A.19(c). The hypotenuse of this triangle is of course the apparent power seen earlier. The apparent power S in VA can be visualized as a complex quantity with a real component $P = VI \cos\theta$ in watts and an imaginary component $Q = VI \sin\theta$ in VAR. The complex power S is therefore defined as

$$S = VI\cos\theta + jVI\sin\theta = P + jQ \tag{A.27}$$

Note that the power factor angle θ is the same angle in all three triangles.

Now the complex power is ready to be derived directly from the applied voltage and the resulting current in a circuit element. In the element of Figure A.20 the phasors of voltage and current can be expressed in general by the exponential form

$$V_{AB} = V_{AB}e^{j\alpha} \qquad \text{and} \qquad I_{AB} = I_{AB}e^{j\beta}$$

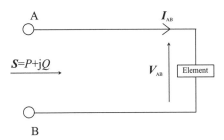

Figure A.20   Reference directions for *V, I, S, P and Q*

where α and β are the angles by which V_{AB} and I_{AB} lead a reference phasor. Assume that $\alpha > \beta$. The angle by which V_{AB} leads I_{AB} (the power factor angle) is $(\alpha - \beta)$.

It would seem sensible that the multiplication of these two phasors gives the complex power. This task will now be carried out:

$$V_{AB}e^{j\alpha}I_{AB}e^{jb} = V_{AB}I_{AB}e^{j(\alpha+\beta)} = V_{AB}I_{AB}\cos(\alpha+\beta) + jV_{AB}I_{AB}\sin(\alpha+\beta)$$

Unfortunately, this product gives the wrong result! In fact, there is no good reason why an algebraic manipulation would invariably produce a result that complies with the adopted conventions. An alternative product will be tried using the *conjugate* of the current phasor I_{AB}^* in which the sign of the angle is negated:

$$S = V_{AB}I_{AB}^* = V_{AB}e^{j\alpha}I_{AB}e^{-jb} = V_{AB}I_{AB}e^{j(\alpha-\beta)} = V_{AB}I_{AB}\cos(\alpha-\beta) + jV_{AB}I_{AB}\sin(\alpha-\beta)$$

This gives the correct result; hence it can be confirmed that in general

$$S = VI^* = P + jQ \tag{A.28}$$

The accepted reference directions of voltage, current, P and Q are shown in Figure A.20, which depicts any power network element. If the complex product of V_{AB} and I_{AB}^* results in positive P and Q (i.e. the active and reactive power flow into the element terminals), the element is a consumer of P and Q and operates in the first quadrant of Figure A.18. In Chapters 4, 5 and 6 it is shown that, depending on its nature, an element could operate in any of the four quadrants.

A.13 Conservation of Active and Reactive Power

Unlike voltage and current in AC systems, active power is a scalar quantity. If a heating element and an induction motor are connected at the terminals of a consumer, the total active power absorbed from the mains is the scalar sum of the two active powers associated with the two components. This is, of course, demanded by the conservation of energy principle.

By analogy to active power conservation, a similar notion applies to reactive power. An element consisting of an inductor and a capacitor absorbs a total reactive power equal to the sum of the component reactive powers. If the Q of the inductor is larger than the Q of the capacitor, the grid supplies reactive power to the *L–C* combination. If the capacitive Q is larger than the inductive Q, reactive power is supplied to the grid. In the first case the L–C combination is a net consumer of Q, in the second case it is a net generator.

These considerations are summarized by the following equations, which apply to any node, bounded portion or the totality of a power system network under steady state conditions:

$$
\begin{aligned}
P_{entering} - P_{leaving} - P_{loss} &= 0 \Rightarrow \sum P = 0 \\
Q_{entering} - Q_{leaving} - Q_{loss} &= 0 \Rightarrow \sum Q = 0
\end{aligned}
\tag{A.29}
$$

The P_{loss} and Q_{loss} refer to any consumption of P and Q in the transmission system (see the next section).

In Chapter 5, it is shown that the concepts underlying Equations (A.29) are vital in the calculation of power flows in power system networks.

A.14 Effects of Reactive Power Flow – Power Factor Correction

Figure A.21 represents a basic circuit of energy transportation from a generator to a consumer through a transmission line, which is simply represented by a series inductance and resistance. The consumer is represented by an inductive resistive impedance. The reason for these representations are explained in Chapter 5.

The consumer 'absorbs' both P_L and Q_L. The generator or 'the utility' has to supply both the P_L of the load and the transmission loss given by I^2R_t, where R_t is the resistance of the transmission line. It also has to supply the Q_L of the load plus the I^2X_t absorbed by the reactance of the transmission line. The transmission loss is therefore strongly influenced by the transfer of Q_L.

To appreciate this, assume that the load consists solely of an inductor, which absorbs Q but not P. In this case the energy meter at the consumer premises, which records kilowatt-hours (i.e. energy purchased), indicates zero but the finite current in the transmission system results in I^2R_t, which has to be supplied by the utility. Clearly this is a most undesirable scenario which the utility endeavours to discourage, in the case of large consumers, through special tariffs that penalize the absorption of Q.

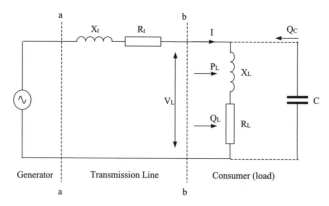

Figure A.21   Simple transmission system

Additionally, the utility has the legal obligation to supply power to the consumer at a more or less fixed voltage. In Chapter 5 it is shown that for transmission lines, with $X_t > R_t$, the voltage V_L at the consumer terminals is particularly sensitive to changes in Q rather than P. It should now be obvious that the utility encourages consumers to draw P at minimum Q, i.e. with $\cos \phi$ as close to unity as possible.

If a capacitor C were to be connected across the load terminals and sized to generate Q_C so that $Q_C = Q_L$, then the consumer will appear to the utility as having a unity power factor. The utility would be delighted with such an arrangement as this will minimize losses in the transmission line resistance R_t and will ensure minimal variation in consumer voltage. This procedure is known as *power factor correction*. Such capacitors are used extensively in power systems to generate or 'inject' reactive power at strategic points of the network.

This discussion also highlights why power system equipment is rated in terms of apparent power *VA* rather than P or Q. As an example take the transmission line in Figure A.21. This line is suspended from insulators that have been designed to withstand (including a safety factor) the nominal voltage of transmission V_L. It is also made of a conductor that has a specific resistance per km and chosen so that when the rated current is carried the heat generated does not exceed a level at which the expansion of the line produces an illegal sag at the lowest point of the catenary. With a purely inductive consumer load absorbing the rated current at nominal voltage the transmission line will be fully loaded in spite of the fact that it transports no useful power. Similar considerations apply to other power system equipment.

A.15 Three-phase AC

Power systems are almost universally three-phase AC. This gives a significant saving in the construction of all power system equipment from generators to transmission lines and consumer equipment compared to single-phase [1]. Even more importantly, three-phase very elegantly provides the rotating magnetic field required in synchronous and induction generators, as described in Chapter 4.

The three windings of the generator in Figure 4.2 are shown again in Figure A.22, feeding individual consumers by means of six independent wires. The orientation of the symbols at $120°$ intervals mimics the orientation of the respective voltage phasors. The voltage V_{ph} is a *phase voltage*, and the same voltage magnitude could be measured on either of the two other phases. Such a system would be perfectly capable of producing a rotating magnetic field, but the use of six wires is highly inefficient.

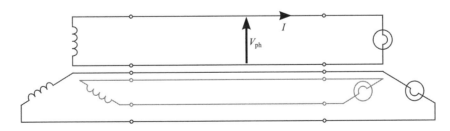

Figure A.22 Three-phase with six wires – three separate circuits

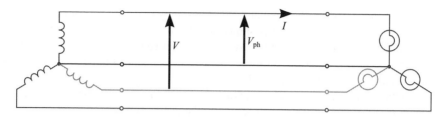

Figure A.23   Three-phase with four wires

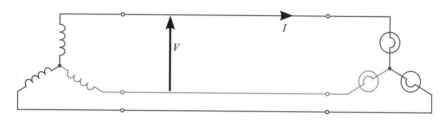

Figure A.24   Three-phase with three wires

In Figure A.23 the three conductors bunched together in Figure A.22 are merged into one common return conductor known as the neutral. With the circuits combined in this way, a voltage can now be measured between any two outer lines. This is known as the line-to-line voltage or simply the *line voltage* and magnitude-wise is the same no matter which two lines are chosen.

When discussing three-phase electricity it is normal practice to quote line voltages rather than phase voltages. Thus, the line voltage in Figure A.23 is indicated by V (without a subscript). Consideration of the geometry of the related phasor diagram shows that (under balanced conditions)

$$V = \sqrt{3}V_{ph}$$

where V is the line voltage and V_{ph} is the phase voltage. The four-wire arrangement is very widely used at the low voltage distribution part of a power system (Figure 1.12); for example $V = 400\,\text{V}$ and $V_{ph} = 230\,\text{V}$.

When a three-phase system is perfectly balanced, the currents in the three phases have equal magnitudes and angles at exactly 120° intervals. Thus, the phasor sum of these currents is zero and so the current in the neutral is also zero. In this case, the neutral can be removed and the system operated with just three wires as shown in Figure A.24.

In practice, perfect balance of the three phases is not an absolute requirement. A three-wire system can operate in an unbalanced condition, but this is usually undesirable and is certainly harder to comprehend. Unbalanced operation is mentioned where necessary, but in general balanced operation is normally assumed in this book. The three-wire arrangement shown is very widely used at 'higher' voltages, i.e. 11 kV and above.

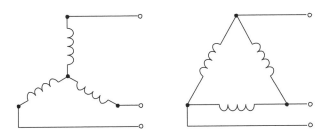

Figure A.25   Star and delta connections

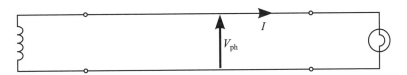

Figure A.26   Single-phase circuit used for calculations

The three windings of a three-phase generator can be connected either in *star* or *delta*, as shown in Figure A.25. Three-phase loads and transformers are similarly configured, and it is normal to have a mixture of star-connected and delta-connected equipment in any three-phase system.

The star configuration can provide a neutral connection for use in a four-wire system. The delta has only three external connections. There are subtle pros and cons associated with both star-connected and delta-connected configurations, particularly regarding unbalanced and faulted conditions, but these matters are beyond the terms of reference of this book.

So long as the three phases are perfectly balanced, it is perfectly legitimate and accurate to carry out calculations by considering just one of the three phases, taking care to include factors of 3 and root-3 appropriately. With this in mind, the three-phase circuit diagram of Figure A.24 can be reduced to a single phase, as shown in Figure A.26.

It must be observed that the power in each phase is one-third of the total power, and, as noted above, that $V = \sqrt{3}V_{\text{ph}}$. Even in a three-wire system or where equipment is internally connected in delta, it is perfectly legitimate and accurate to use the above approach: simply imagine a neutral conductor and reference voltages to it. Notice also that the current I is the same in all the preceding diagrams: there is no distinction between line current and phase current in this context. Contrast this with the currents in a delta, where $I_{\text{line}} = \sqrt{3}I_{\text{ph}}$.

A.16 The Thévenin Equivalent Circuit

The one-port circuit of Figure A.1 was initially assumed to be *passive*, i.e consisting of R, L or C components. In contrast, *active* networks include energy sources. A valuable method of representing active networks by simpler equivalent circuits is based on the Thévenin theorem. This theorem states that any one-port network consisting of passive elements and energy

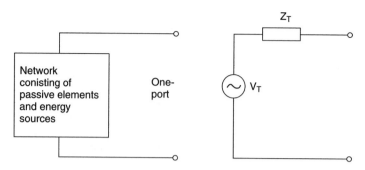

Figure A.27   Thévenin equivalent circuit

sources can be replaced by an ideal voltage source V_T (the Thévenin voltage) in series with an impedance Z_T (the Thévenin impedance). V_T is the open-circuit voltage of the two-port and Z_T is the ratio of the open-circuit voltage divided by the short-circuit current. This equivalence, which is of considerable value in power system analysis, is illustrated in Figure A.27. Proof of this theorem can be found in Reference [1] or any other text on electric network theory.

Reference

[1] Smith, R.J. and Dorf, R.C. *Circuits, Devices and Systems*, John Wiley & Sons, Ltd, Chichester, 1992.

Index